Risk Analysis in Building Fire Safety Engineering

Risk Analysis in Building Fire Safety Engineering

A.M. Hasofer
V.R. Beck
I.D. Bennetts

ELSEVIER

AMSTERDAM • BOSTON • HEIDELBERG • LONDON • NEW YORK • OXFORD
PARIS • SAN DIEGO • SAN FRANCISCO • SINGAPORE • SYDNEY • TOKYO
Butterworth-Heinemann is an imprint of Elsevier

Butterworth-Heinemann is an imprint of Elsevier
Linacre House, Jordan Hill, Oxford OX2 8DP, UK
30 Corporate Drive, Suite 400, Burlington, MA 01803, USA

First edition 2007

Notice
No responsibility is assumed by the publisher for any injury and/or damage to persons
or property as a matter of products liability, negligence or otherwise, or from any use
or operation of any methods, products, instructions or ideas contained in the material
herein. Because of rapid advances in the medical sciences, in particular, independent
verification of diagnoses and drug dosages should be made

British Library Cataloguing in Publication Data
A catalogue record for this book is available from the British Library

Library of Congress Cataloging in Publication Data
A catalog record for this book is available from the Library of Congress

ISBN-13: 978-0-750-68156-8
ISBN-10: 0-7506-8156-X

For information on all Butterworth-Heinemann publications
visit our website at books.elsevier.com

Typeset by Charon Tec Ltd (A Macmillan Company), Chennai, India
www.charontec.com
Printed and bound in Great Britain

07 08 09 10   10 9 8 7 6 5 4 3 2 1

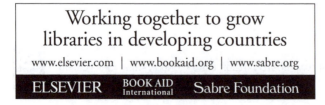

Contents

Dedication

With the consent of my co-authors, this book is dedicated to my wife Renée Hasofer, without whose patience, encouragement, dedication and love this book would never have been written.

A.M. Hasofer

1

Introduction

Society has responded to the threat of fire in buildings in many ways, including: fire brigades, insurance, building regulations, education on fire hazards, controls on the use of materials and products in buildings and the design of buildings to resist the effects of fire. The level of fire safety and protection in buildings reflects the general economic, social and cultural features of society.

Building regulations are an important component in the design for fire safety and protection in buildings. The prescriptive design requirements in building regulations were introduced many years ago and were applicable to the technology and practices then in vogue. Many of the regulatory provisions were empirically derived, but they have assumed great authority with the passage of time, although perhaps lacking technical substantiation. These prescriptive requirements in building regulations reflect the low level of technology previously available for the design of fire safety and protection in buildings.

It is generally acknowledged that the current design approach specified in building regulations has resulted in the achievement of fire safety and protection levels that the community appears to accept. However, the prescriptive design approach is unlikely to result in the most cost-effective design solutions, nor in designs which maintain a consistent level of safety and protection in buildings. Furthermore, the prescriptive regulatory design approach restricts the range of design choices that are available; this inhibits and restricts innovation (Warren Centre Project Report in Ref. [23] pp. 7, 8).

In recent decades, a more systematic engineering approach has been developed for the design of fire safety and protection. This has followed from a more scientific understanding of the fundamental aspects of fire initiation, growth and spread in buildings, and the response of materials, structures and people to the effects of fire. The fire engineering approach provides opportunities for reducing the overall cost of providing fire safety and protection measures in buildings and to introduce greater flexibility to design. However, with the traditional fire safety engineering approach, designs are normally implemented without an explicit determination of fire safety and protection levels. Traditional fire safety engineering is normally performed by evaluating designs according to a specified fire scenario. In traditional fire safety engineering, the design solution arising from the prescriptive regulatory method is normally used as a starting point and modifications are then made to the prescriptive design; these are normally called "trade-offs". The approach to "trade-offs" is relatively simple but constrained. Normally, one fire safety system is changed or deleted while another is added or modified with the intent of achieving the same level of performance, whether this is safety or protection. However, the fire engineering approach using the "trade-off" method does not normally explicitly consider safety or protection levels.

If designers are to explicitly consider the levels of fire safety and protection in buildings designs, then there must be a fundamental change in the methods that are used to design buildings for fire safety and protection. In addition, there must be recognition that societal levels of fire safety and protection in buildings are the result of a multitudinous number of fire scenarios, multiple responses of building fire safety and protection subsystems to these fire scenarios and a multitudinous number of human behaviour responses to these scenarios. A rigorous and systematic approach to the assessment of explicit levels of fire safety and protection requires a comprehensive risk analysis to be undertaken of building fire safety and protection systems. Fundamentally, this requires explicit consideration of multiple fire scenarios, response of building fire safety systems and human behavioural responses.

Risk assessment models are needed to identify those combinations of building subsystems which provide the requisite level of safety in a cost-effective manner. Deterministic fire-engineering design methods cannot be used for that purpose, because it is necessary to estimate both the likelihood of the possible fire scenarios and their consequences, and then combine the results in order to evaluate the likely cost and the likely safety level.

Risk analysis is defined as the process of estimating magnitudes of consequences and probabilities of the adverse effects resulting from fire in a building. It provides rational criteria for the choice of remedial actions, including explicit consideration of uncertainty. It is obviously the preferred base for decision making.

Research into building fire safety using a risk-based engineering methodology to estimate both risk to life safety and fire cost was first undertaken by Prof. V.R. Beck in 1979. It was further carried out at the Warren Centre for Advanced Engineering, University of Sydney, Australia. Previous research into the effects of fire in buildings had been conducted in disparate areas such as fire initiation and growth, smoke spread, fire spread, fire extinguishment and smoke control facilities, effects of fire on building components and structure and human response to fire. Beck was the first to develop an integrative methodology that enabled each of these areas of research to be combined in a way that yielded estimates for both "risk to life safety" and "the expected fire cost" (cost of fire protection and expected fire losses) associated with the effects of fire for a given building design. These performance parameters can be used to identify cost-effective fire safety design solutions for buildings. Accordingly, for the first time a systematic approach was available to translate building design and occupancy into a quantitative measure of life safety. A distinguishing feature of Beck's work has been the representation of the complex interactions between fire, buildings and people using a temporal risk assessment methodology. His work has enabled the results of previous research to be combined and used, and he has identified the need and provided a focus for substantial new research to support the systematic quantification of fire safety performance in buildings. This research has brought a new understanding to building fire. It has served to focus the interests of industry, government and the community into a common research theme which has benefited all three groups.

There is currently no book that covers the topic of Performance Based Design for Building Fire Safety based on Risk Assessment. It is that gap that this book attempts to bridge. It aims at providing a self-contained presentation of Risk Analysis in Building Fire Safety Engineering suitable for courses and/or self-study, with minimal mathematical prerequisites. It is an important contribution to the understanding of comprehensive analysis methods to predict the levels of fire safety and protection in buildings. It draws on results in risk analysis that have been published over the last 27 years as Internal Reports or in article form in the main Fire Engineering Journals and main International Conferences. It brings together for the first time a comprehensive treatise on the subject of risk assessment of fire safety and protection in buildings.

The book requires no prerequisites in Probability and Statistics or Fire Engineering. All required knowledge in these areas is included in the preliminary chapters. All examples used in these chapters are drawn from Fire Engineering. The book is suitable for teaching Fire Engineering components of Engineering Courses at the Senior Undergraduate level, Fire Engineering Postgraduate Courses and Refresher Courses for Fire Engineers. In addition, anyone involved in Fire Engineering and/or Risk Analysis in Buildings would significantly benefit from studying the book and having it available for reference. Prerequisites do not exceed knowledge of elementary mathematics and physics at the level of a science-oriented First Year University.

Chapter 2 contains a general description of fires in buildings. It describes the combustion phenomenon, enclosure fires and the extent of their spread, and the fire safety system.

Chapter 3 is a presentation of the elements of probability theory required for the understanding of risk analysis. It covers events and probability, random variables, expectation, elements of joint distributions, estimation and confidence intervals.

In Chapter 4, a description is given of the beta reliability index, a measure of system safety that has become increasingly popular in many fields of engineering.

In Chapter 5, the Monte Carlo Method is presented. This provides a systematic approach to situations in risk analysis when there is no practical analytic algorithm to evaluate the required probabilities.

Event and fault trees are discussed in Chapter 6. They are a powerful tool to identify failure scenarios and evaluate their probabilities.

Chapter 7 provides a description of implementing performance-based optimal design solutions based on expected cost–benefit analysis.

Several chapters then follow on modelling the probabilistic and stochastic aspects of various fire safety subsystems.

Chapter 8 contains a risk analysis of fire initiation. It covers ignition frequency and the probability distribution of fire losses. The most significant fire factors resulting in significant damage to life are identified using the Fire Incident Data of US Fire Statistics.

In Chapter 9, the personal factors that result in significant damage to life are identified, using the Civilian Casualty Data of US Fire Statistics.

In Chapter 10, a probabilistic modelling of barrier resistance for wood-framed wall construction is presented. It enables the evaluation of a cumulative probability of failure with time curve as part of an overall optimal performance-based design.

Chapter 11 contains a description of a stochastic fire growth model in an apartment unit. It illustrates a methodology for converting a deterministic model to a stochastic model and then goes on to use quadratic response surfaces, the beta reliability index and Monte Carlo simulation to study the maximum temperature reached and the time to untenable conditions.

Chapter 12 contains a description of a stochastic model of smoke spread, illustrating a different method for converting a deterministic model into a stochastic model. It then goes on to use Monte Carlo simulation to show how to reconcile the model with experimental results.

Given in Chapter 13 is a stochastic model of human behaviour that covers occupant response and evacuation.

Chapter 14 deals with the performance assessment of fire safety systems, using fire statistics. Protected construction, detectors and sprinklers are studied.

Chapter 15 contains a description of stochastic modelling of fire brigade response. It includes models for travel time, frequency of blockage and fire extinction performance.

Following the descriptions of the stochastic models of fire safety systems, the focus is then to illustrate the performance assessment of fire safety systems and to predict the levels of fire safety and protection in buildings. Two case studies are given of designs of performance-based fire safety systems.

Chapter 16 illustrates the performance-based safety design of an assembly hall, while Chapter 17 presents the case history of a 41-storey building at 140, William Street, Melbourne, Australia. It vividly illustrates the potential benefits of risk-based fire safety design.

2

General description of building fires

2.1 Introduction

The development and management of fires in buildings is often seen as a deterministic process whereby particular inputs result in well-defined outputs. However, it is better to recognize that there are uncertainties associated with both the prediction of fire development and the performance of the various measures employed to manage the effects of fires. The presence of such uncertainty implies that there is a probability that the fire safety design objectives may not be achieved for a given building. The process of risk assessment is therefore fundamental to deciding which building design is better from a fire safety viewpoint, and given a particular building design, the most effective on-going management protocol.

2.2 The combustion phenomenon

Fire is defined primarily as rapid oxidation accompanied by heat and light. In general, oxidation is the chemical union of any substance with oxygen. The rusting of iron is oxidation but it is not fire because it is not accompanied by light. Heat is generated, but so little that it can hardly be measured. Burning can occur as a form of chemical union with chlorine and some other gases, but for our purpose we need to only consider fire that involves oxygen.

2.2.1 The classic triangle concept of fire

Fire can usually take place only when three things are present: oxygen in some form, fuel (material) to combine with the oxygen and heat sufficient to maintain combustion. Removal of any one of these three factors will result in the extinguishment of fire. The classic "fire triangle" (see Figure 2.1) is a graphical symbolization of the recognized elements involved in the combustion process. Opening the triangle by removing one factor will extinguish a growing fire, and keeping any one factor from joining the other two will prevent a fire from starting.

2.2.2 The tetrahedron concept of fire

Recent research suggests that the chemical reaction involved in fire is not as simple as the triangle indicates and that a fourth factor is present. This fourth factor is a reaction chain where burning continues and even accelerates, once it has begun.

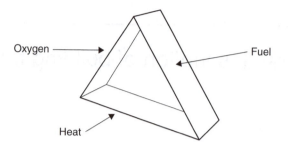

Figure 2.1 Fire triangle.

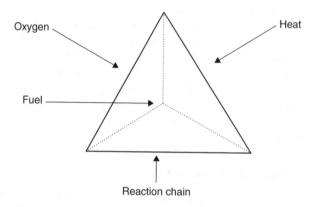

Figure 2.2 Fire tetrahedron.

Haessler [26], in his study of fire, formulated the theory of the diffusion flame combustion phenomenon as a tetrahedron. Haessler preferred to symbolize his concept of fire as a tetrahedron instead of a square because in the tetrahedron the four entities are adjoining and each is connected with the other three entities.

This reaction chain is caused by the breakdown and recombination of the molecules that make up a combustible material with the oxygen of the atmosphere. A piece of paper, made up of cellulose molecules, is a good example of a combustible material. Those molecules that are close to the heat source begin to vibrate at an enormously increased rate and, almost instantaneously, begin to break apart. In a series of chemical reactions, these fragments continue to break up, producing free carbon and hydrogen that combine with the oxygen in the air. This combination releases additional energy. Some of the released energy breaks up still more cellulose molecules, releasing more free carbon and hydrogen, which, in turn, combine with more oxygen, releasing more energy and so on. The flames will continue until fuel is exhausted, oxygen is excluded in some way, heat is dissipated or the flame reaction chain is disrupted.

Investigation of this concept has led to the discovery of many extinguishing agents that are more effective than those that simply manage to open the triangle. Because of this discovery, we must modify our fire triangle into a three-dimensional pyramid, known as the "tetrahedron of fire" (see Figure 2.2).

This modification does not eliminate old procedures in dealing with fire but it does provide additional means by which fire may be prevented or extinguished.

2.3 Fire spread

The rate at which fire will develop will depend on how rapidly flame can spread from the point of ignition to involve an increasingly large area of combustible material. Flame spread is considered as an advancing ignition front in which the leading edge of the flame acts both as the source of heat, to raise the fuel ahead of the flame front to the fire point, and as the source of pilot ignition. There are various factors which are known to be significant in determining the rate of flame spread over combustible solids: material factors and environmental factors.

Environmental factors consist of composition of atmosphere, temperature, imposed heat flux and air velocity. Composition of the atmosphere refers to the oxygen concentration. Combustible materials will ignite more readily, spread flame more rapidly and burn more vigorously if the oxygen concentration is increased. Higher rates of flame spread are observed with effective oxygen enrichment which enhances flame stability at the surface. Temperature refers to the temperature of the fuel. Increasing the temperature of the fuel increases the rate of flame spread, the higher the initial fuel temperature the less heat required to raise the unaffected fuel to the fire point ahead of the flame. An imposed radiant heat flux causes an increase in the rate of flame spread, by preheating the fuel ahead of the flame front. Confluent air movement enhances the rate of flame spread over a combustible surface. Friedman [24] reports that the rate will increase quasi-exponentially up to a critical level at which extinction will occur.

Material factors are further divided into chemical and physical factors. The chemical factors consist of composition of fuel and presence of retardants. The physical factors consist of initial temperature, surface orientation, direction of propagation, thickness, thermal capacity, thermal conductivity, density, geometry and continuity. As an example, of surface orientation effect, Alpert and Ward [5] point out that the spread of a flame along a vertical surface accelerates exponentially.

The development of fires within buildings is very complicated given the variety of potential ignition sites, materials, geometry and ventilation. Fire statistics relating to the extent of fire spread provide some interesting findings from an overall perspective. Fire statistics are collected by the fire brigade and record the extent of flame spread in terms of the following categories:

- Restricted to the object of fire origin
- Restricted to the area of fire origin
- Restricted to the room of fire origin
- Restricted to the compartment of fire origin (implies a fire-resistant enclosure)
- Restricted to the floor of fire origin
- Restricted to the building of fire origin.

Table 2.1 Extent of flame spread for apartments and offices.

Extent of flame spread	Apartment buildings (%)	Commercial buildings (%)
Confined to object of origin	36.0	41.0
Confined to area of origin	25.0	19.6
Confined to room of origin	18.0	13.9
Confined to fire-rated compartment of origin	0.7	0.5
Confined to the floor of origin	2.3	2.0
Confined to structure of origin	15.0	19.8
Extended beyond the structure of origin	2.0	3.2

Table 2.1 (from Fire Code Reform Centre [13]) reveals the following fire spread characteristics for unsprinklered office and apartment buildings: the majority of fires do not extend beyond the room of fire origin; however, it appears that of those that extend beyond the room to the compartment of origin, only relatively few are contained within the compartment. The table is based on an analysis of US statistics for the years 1983–1991. These trends are matched closely by statistics from other countries.

As far as office buildings are concerned, the time of day has an important effect. There are more fire starts during normal operating hours but less larger fires. This is shown in Figure 2.3, again for unsprinklered US office buildings. The presence of people within the building results in more electrical equipment operating and more human activity, but the presence of a greater number of occupants means that it is much less likely that fires will be allowed to spread. On the average, it is 6 times less likely that a fire will spread beyond the room of origin if the fire occurs during normal operating hours.

A more detailed analysis of such statistics and their relevance to the development of flame spread models will be presented in Chapter 8.

2.4 Enclosure fires

Buildings are conveniently considered as being divided into various enclosures: rooms, combination of rooms, storey. The term "compartment" usually refers to an enclosure which has been designed to "contain" the fire. It might therefore refer to a room with fire-resistant walls or a larger space containing rooms but where fire-resistant walls bound this space.

As illustrated by the previous statistics, a fire can potentially grow beyond the object to the area of fire origin, then to the room and finally to the compartment and beyond. The various stages of such a fire can be schematically represented in the manner represented by Figure 2.4.

The various stages are said to be:

- Ignition
- Growth
- Fully developed
- Decay.

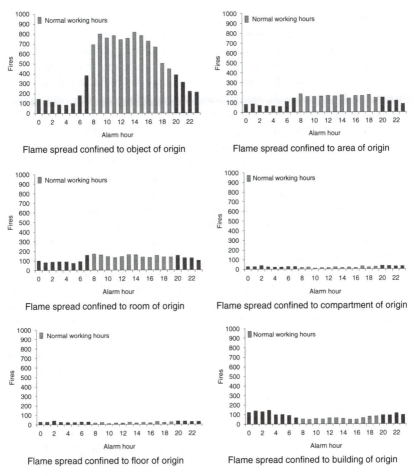

Figure 2.3 Time of day dependence of flame spread.

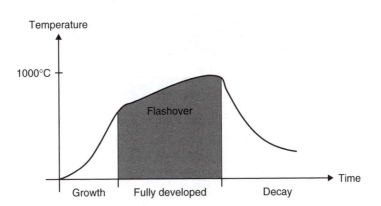

Figure 2.4 Graph of the change in temperature over time in an enclosure fire.

The transition from the growth to the fully developed fire stage is defined as "flashover". Walton and Thomas [69] defined a flashover as the "transition from a growing fire to a fully developed fire within the compartment in which all combustible items are involved in the fire". In small cubic shaped enclosures (e.g. 3 m × 3 m × 3 m), all of the fuel surfaces may be simultaneously involved in the burning process with, more or less, uniform gas temperatures throughout. For other shape enclosures, or larger enclosures, the picture is not likely to be as simple as this. During the fully developed stage, burning may only occur in parts of the enclosure and there may be significant variations in gas temperature.

In the event of a fire within an enclosure, oxygen to support combustion comes from the air within the enclosure and from that drawn in through openings in the enclosure boundaries. It has been long recognized that the area (A) and height (h) of openings in enclosure boundaries have a dominant influence on the rate of burning within the enclosure. This is true for situations where combustion is controlled by the flow of air into the enclosure. For small cubic shaped enclosures with single openings, the rate of burning is correlated with $A\sqrt{h}$. However, the geometry of the enclosure and the associated position of openings will influence the flow of gases within the enclosure and therefore directly influence the rate of burning. This is so is illustrated by the following test results for three enclosures having exactly the same geometry but where the openings are in different locations (see Figure 2.5).

In each case the total $A\sqrt{h}$ is the same where A is the area of an opening and h is the corresponding opening height. In the first case the opening is at the front of the enclosure; in the second, at the centre of the side; and in the third case, an opening at the front and rear with front opening in the upper half of the enclosure and the rear opening in the lower half. The tests were conducted using ethanol in steel trays equally distributed around the floor of each enclosure.

The comparative rates of burning for each of the cases are given. If wood, instead of ethanol, is used, the same result is obtained. The flow of gas within each of the enclosures is significantly different. In the first case, the flows are

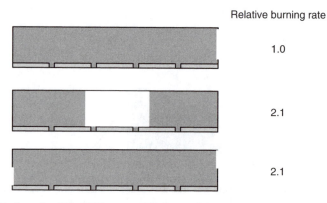

Relative burning rate

1.0

2.1

2.1

Each enclosure is 3000 mm × 600 mm × 1200 mm

Figure 2.5 Effect of opening location on burning rate.

limited to the front of the enclosure (where the opening is located) where burning takes place preventing oxygen getting to the fuel behind the flame front. The fire moves to the next set of trays only when the fuel in the front trays has gone. In the second case highly turbulent flows are set up on each side of the opening and the volatiles from trays on both sides and adjacent to the opening are entrained into the flows and burnt at the opening. In the final case, air is drawn from the low level vent and a longitudinal flow is set up whereby air is supplied to the burning trays and burning gases are exhausted at the high level opening.

All fires manifest an ignition stage but, beyond that, may fail to grow through all or some of the growth stages listed.

2.5 The fire safety system

2.5.1 Introduction

There are many aspects within buildings that affect fire safety. It is convenient to consider those aspects that reduce the likelihood or consequences of fires as constituting a fire safety system. Each of these aspects is then considered as a subsystem. Such a conceptual framework is essential to allow the systematic risk modelling of fire safety within a building.

A fire safety system can be considered to have the following objectives in relation to a building considered as a collection of enclosures, which may or may not have fire-resistant boundaries:

1. Within the enclosure where the fire is initiated:
   (a) control of fire initiation and development in the early stages,
   (b) provide early warning and suitable evacuation paths for occupants within the enclosure.
2. Outside the enclosure where the fire is initiated:
   (a) control the spread of flame,
   (b) limit the spread of smoke to minimize impact on occupants outside the enclosure of fire origin,
   (c) provide occupants with early warning and evacuation paths that are sufficiently smoke and heat free to allow them to avoid the effects of the fire,
   (d) provide sufficient structural stability for the occupants to escape and allow reasonable fire brigade activity.

2.5.2 Subsystems

Introduction

The various subsystems can be classified according to the above objectives or functions:

1. Control of fire ignition and development in early stages
2. Control of flame spread
3. Control of spread of smoke and toxic products

Table 2.2 Summary of measures for various subsystems

Subsystem	Possible measures	
	"Hardware"	"Software"
Control of fire initiation and development in early stages	• Earth leakage devices • Surveillance systems • Materials of construction • Alarm and detection systems plus hose reels and extinguishers • Sprinklers • Other automatic fire suppression hardware	• Regular maintenance of electrical and mechanical systems • Human monitoring of surveillance systems • Presence of occupants within the building • Presence of occupants trained in early fire fighting in building • Management and maintenance of alarm and detection systems • Maintenance of hose reels and extinguishers • Management and maintenance of sprinkler systems
Control of flame spread	• Physical barriers • Materials of construction including linings • Alarm and detection systems plus fire brigade	• Maintenance of barriers • Management and maintenance of alarm and detection systems
Control of spread of smoke and toxic products	• Physical barriers • Smoke exhaust systems (purging) • Pressurization systems (e.g. stairs or zones)	• Maintenance of barriers • Management and maintenance of smoke exhaust and pressurization systems
Provision of means to allow occupant avoidance	• Signage • Exits	• Presence of trained wardens • Evacuation drills
Provision of structural adequacy	• Size of structural members • Overall structural behaviour • Fire protective coatings, concrete cover	• Maintenance of coatings

4. Provision of means to allow occupant avoidance
5. Provision of sufficient structural adequacy.

Each of these subsystems must be considered in relation to the possible measures incorporated within buildings aimed at achieving fire safety. These measures can be considered to be composed of "hardware" and "software". The term "hardware" refers to physical systems that are incorporated in buildings such as sprinklers, smoke exhaust or fire-resistant barriers; whereas the term "software" refers to human activities within a building that can have either a direct influence on fire safety or an indirect influence on the reliability and efficacy of hardware within the building. The effectiveness of each subsystem can be taken as some combination of reliability and efficacy. A reliability of unity means that the subsystem will always operate. An efficacy of unity means that the subsystem will perform to the level expected given that it operates. The reliability and efficacy of real subsystems are always less than unity.

Stochastic models for the various subsystems attempt to model the reliability and efficacy of the various subsystems.

Consideration of subsystems

Table 2.2 summarizes the various measures that may be employed within a building to form a given subsystem. Not all of these measures are appropriate for all buildings.

It should be noted that the first subsystem has a significant effect on all of the other subsystems. This is because extinguishing a fire at its early stages may reduce the need for the other subsystems. Studies such as for example Bennetts *et al.* [38] have pointed out that the most effective strategy in controlling the development of a fire is to provide an effective and highly reliable sprinkler system to extinguish the fire in its early stages. Widespread damage and loss of life in large buildings almost invariably occur as a result of failure of measures to control the development of a fire.

The contribution of the various subsystems to overall fire safety can be determined by the application of risk assessment methodology.

3

Elements of probability theory

3.1 Why do we need probability theory?

In many fields of engineering, achieving a specific target is done by evaluating the deterministic performance of the design under study, with possibly a few representative scenarios.

Unfortunately, the phenomena which constitute a building fire and the concomitant fire fighting exhibit much uncontrollable variability. The sources of this variability are, among others:

1. The inherent variability of fire and smoke spread.
2. Lack of knowledge regarding the type, quantity and geometry of the fuel.
3. Lack of knowledge regarding whether doors and windows are closed or open.
4. Lack of knowledge regarding the occupants of the building and their likely reaction to the fire.
5. Lack of knowledge regarding the actions of the fire brigade.

In addition, there are other sources of variability which derive from the shortcomings of the modelling of the fire.

Thus, simulating a fire with a few deterministic models may be very misleading. It is necessary to take into account how often the various possibilities for each event do occur.

3.2 The sample space

Suppose that we conduct an experiment and observe its results. In fire engineering we obtain quite often widely varying results as we repeat the experiment, even though we have tried to keep the conditions constant.

To model this situation formally we introduce a set which we call the *sample space*. A set is a collection of elements. Here the elements of the sample space are all the possible outcomes of the experiment. The sample space is usually denoted by S.

For example, if we light a fire in a compartment, we could consider three possible outcomes: a smouldering fire, a flaming fire and a flashover fire. In that case, the sample space of the experiment consists of three elements. If we repeat the experiment often enough we may be able to state that we get a smouldering fire 30% of the time, a flaming fire 50% of the time and a flashover fire 20% of the time. In this way, we can attach *probabilities* to the elements of the sample space. Probabilities are expressed as fractions, not percentages. Here the three outcomes will be said to have probabilities of 0.3, 0.5 and 0.2, respectively.

Sometimes there is an infinity of possible outcomes. For example, we might be measuring the maximum temperature reached in a compartment. It might be any number between, say, 500°C and 1500°C. The sample space in this case is the interval (500, 1500).

3.2.1 Events

An *event* is a subset of the sample space. For example, we may be interested in fires where the maximum temperature reached is between 1000°C and 1200°C. The interval (1000, 1200), a subset of the sample space (500, 1500), is called an *event*.

Definition 3.2.1. *An event that consists of only one point of the sample space, i.e. just one outcome, is called an* elementary event.

Since events are sets, they obey the laws of set operations. Let A and B be two events. Then:

1. The *complement* of A, A^C, also called not A, is the set of elements of S not in A.
2. The *union* of A and B, $A \cup B$, also called A or B, is the set of elements of S in either A or B (i.e. the symbol $\cup$ stands for *or*).
3. The *intersection* of A and B, $A \cap B$, also called A and B, is the set of elements of S in both A and B (i.e. the symbol $\cap$ stands for *and*).

Clearly, $A \cup B$ and $B \cup A$ represent the same event. Similarly for $A \cap B$ and $B \cap A$.

A convenient graphical representation of events is the *Venn diagram* where the sample space is represented by some region, usually a rectangle, and events as smaller regions inside the sample space, e.g. circles. Figures 3.1–3.3 illustrate union, intersection and complementation, respectively.

The null set, usually denoted by ϕ, is a set that contains no elements.

Two sets whose intersection is the null set, i.e. two sets which have no common elements, are said to be *mutually exclusive*.

One important concept in set operations is that of a *partition*. A family of sets $\{A_i, i = 1, \ldots, n\}$, where n may be infinite, is said to form a *partition* of the sample space S if:

1. $A_i \cap A_j = \phi$ for all $i \neq j$, i.e. the sets $\{A_i\}$ are mutually exclusive (no elements are common to two sets).
2. $\cup_i A_i = S$, i.e. the union of the $\{A_i\}$ covers the whole sample space.

For example, consider the last example in the last section, where the sample space was the interval (500, 1500) degrees Celsius. Let the temperature be denoted by T. Then the sets $(500 < T \leq 800)$, $(800 < T \leq 1200)$ and $(1200 < T \leq 1500)$ form a partition of the sample space. This means that every element of the sample space belongs to one and only one of the sets that form the partition.

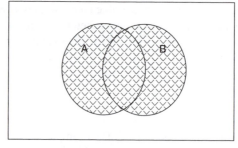

A or B

Figure 3.1 Venn diagram for union.

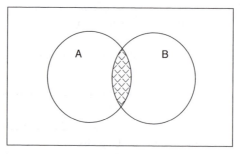

A and B

Figure 3.2 Venn diagram for intersection.

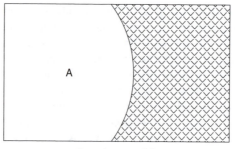

Complement of A

Figure 3.3 Venn diagram for complement.

3.3 The probability measure

To create a mathematical model which will enable us to quantify the likelihood of the various outcomes of the experiment, we attach to each event a number between 0 and 1 which will be called the *probability* of the event. It can be thought of as the relative frequency of occurrence of the event in a large number of repetitions of the experiment. The probability of event A will be denoted by $P(A)$. The probability measure satisfies the following axioms:

1. $0 \le P(A) \le 1$ for all $P(A)$.
2. $P(S) = 1$ and $P(\phi) = 0$.
3. If $A \cup B = \phi$ (i.e. A and B are mutually exclusive), then $P(A \cup B) = P(A) + P(B)$.

It follows immediately from the axioms that if $\{A_1, A_2, \ldots, A_n\}$ is a partition then

$$P(A_1) + P(A_2) + \cdots + P(A_n) = 1. \tag{3.1}$$

Examples

1. Consider a room of fire origin with one door and one window. There are four possible situations, namely: door and window closed, door and window open, door open and window closed, and door closed and window open. By analysing observations on actual fires, we may come to the conclusion that the relative frequencies of these four events are 0.1, 0.2, 0.4 and 0.3.
2. Returning to the maximum temperature T in a compartment, we could conclude from observations, or from computer simulations, for example, that $P(500 < T \le 800) = 0.2$, $P(800 < T \le 1200) = 0.6$ and $P(1200 < T \le 1500) = 0.2$.
3. Let N be the number of fires in 1 day in some fire district. The sample space here is $\{0, 1, 2, \ldots\}$. We could find that $P(N = 0) = 0.61$, $P(N = 1) = 0.30$ and $P(N = 2) = 0.08$. It then follows that $P(N > 2) = 0.01$ since these events form a partition and the sum of their probabilities must therefore be unity.

The probability measure of a set is represented in the Venn diagram by its area.

The following theorems follow obviously from the Venn diagrams.

Theorem 3.3.1. $P(A^C) = 1 - P(A)$.

Theorem 3.3.2. $P(A) = P(A \cap B) + P(A \cap B^C)$.

This is because B and B^C are mutually exclusive. More generally, consider the following two theorems.

Theorem 3.3.3. *Let $\{B_i, i = 1, \ldots, n\}$ be a partition. Then*

$$P(A) = P(A \cap B_1) + P(A \cap B_2) + \cdots + P(A \cap B_n).$$

Theorem 3.3.4. $P(A \cup B) = P(A) + P(B) - P(A \cap B)$.

The negative term comes from the fact that the intersection of A and B is counted twice in the sum $P(A) + P(B)$.

3.4 Conditional probability

Definition 3.4.1.   *For any two events A and B such that $P(B) \neq 0$, the conditional probability of event A given that event B has occurred is denoted by $P(A|B)$ and is defined by*

$$P(A|B) = \frac{P(A \cap B)}{P(B)}. \tag{3.2}$$

Example   Consider example 1 in Section 3.3. Denote the four possible situations (in an obvious way) by $DCWC$, $DOWO$, $DOWC$ and $DCWO$. Suppose, as before, that the probabilities of these events are 0.1, 0.2, 0.4 and 0.3, respectively. Let us note that in reality there are four elementary events: DC, DO, WC, WO and that we can write $DCWC$ as $DC \cap WC$ and so on.

It then follows from Theorem 3.3.2 that, since clearly $WO = WC^C$,

$$P(DC) = P(DC \cap WC) + P(DC \cap WO) = 0.1 + 0.3 = 0.4. \tag{3.3}$$

Similarly, $P(DO) = 0.2 + 0.4 = 0.6$.

We can now ask: among all the compartments which had the door closed, what was the probability (i.e. relative frequency) of open windows? This is exactly what is meant by the probability of an open window, *given that* the door is closed, namely $P(WO|DC)$. Using the formula of Definition 3.4.1, we obtain:

$$P(WO|DC) = \frac{P(DC \cap WO)}{P(DC)} = \frac{0.3}{0.4} = 0.75. \tag{3.4}$$

3.5 The theorem of total probability

Theorem 3.5.1.   *Let $\{A_i, i = 1, \ldots, n\}$ be a partition. Then for any event B*

$$P(B) = P(B|A_1)P(A_1) + P(B|A_2)P(A_2) + \cdots + P(B|A_n)P(A_n). \tag{3.5}$$

Proof   The theorem follows immediately from Theorem 3.3.3 since $P(B \cap A_i) = P(B|A_i)P(A_i)$ for every i.   □

The importance of the theorem of total probability lies in the fact that we can often easily evaluate the probability of some event conditional on the occurrence of some other events. The theorem then enables us to evaluate the unconditional probability.

Example   Suppose that fires in a compartment can be classified into three types with the given probabilities:

1. F_1: Smouldering fire, $P(F_1) = 0.2$.
2. F_2: Flaming fire, $P(F_2) = 0.4$.
3. F_3: Flashover fire, $P(F_3) = 0.4$.

We are interested to find out the probability of death of an occupant of the compartment. Let us denote the event of death by D and suppose that the probability of death, conditional on the type of fire, is given by the following figures: $P(D|F_1) = 0.01$, $P(D|F_2) = 0.3$ and $P(D|F_3) = 0.699$. Then, using the theorem of total probability, we can evaluate $P(D)$ as follows:

$$P(D) = P(D|F_1)P(F_1) + P(D|F_2)P(F_2) + P(D|F_3)P(F_3) \tag{3.6}$$

$$= 0.01 \times 0.2 + 0.3 \times 0.3 + 0.699 \times 0.4 \tag{3.7}$$

$$= 0.3716. \tag{3.8}$$

3.6 The concept of independence

Intuitively, if event B has no effect on event A we can say that A is independent of B. More precisely, if knowledge that event B has occurred does not affect the probability of A, we say that A is independent of B. This is expressed mathematically as:

$$P(A|B) = P(A). \tag{3.9}$$

Replacing $P(A|B)$ by its expression from Definition 3.4.1, we find:

$$\frac{P(A \cap B)}{P(B)} = P(A). \tag{3.10}$$

This can be rewritten $P(A \cap B) = P(A)P(B)$.
It is interesting to note that the last expression implies that

$$\frac{P(A \cap B)}{P(A)} = P(B). \tag{3.11}$$

In other words, if A is independent of B, then B is independent of A. Thus, we can say that if $P(A \cap B) = P(A)P(B)$ then A and B *are independent*.
This definition can be extended to more than two events. We say that $A_1, A_2, \ldots, A_n$ are independent if

$$P(A_1 \cap A_2 \cap \ldots \cap A_n) = P(A_1)P(A_2)\ldots P(A_n). \tag{3.12}$$

Example

1. Returning to the example in Section 3.4, we can ask whether the event "window open" is independent of the event "door open". We have

$$P(WO) = P(DO \cap WO) + P(DC \cap WO) \tag{3.13}$$

$$= 0.2 + 0.3 \tag{3.14}$$

$$= 0.5. \tag{3.15}$$

But

$$P(WO|DO) = \frac{P(DO \cap WO)}{P(DO)} \tag{3.16}$$

$$= \frac{0.2}{0.6} \tag{3.17}$$

$$= 0.33. \tag{3.18}$$

Thus, we see that in the situation considered knowing that the door is open *decreases* the probability that the window will be open, and the two events are *not* independent.

2. Suppose that we know that the probability that a smoke detector is defective is 0.05. Consider *two* smoke detectors in two separate rooms. We can reasonably assume that the two events "smoke detector 1 is defective" and "smoke detector 2 is defective" are independent, if they have been independently bought and independently installed. We can then state that the probability that *both* smoke detectors are defective is $0.05 \times 0.05 = 0.0025$.

   This result is the key to understanding the principle of *redundancy* in system design. Suppose we install several components to carry out the same function in some system. Suppose further that the components have independent probabilities of failure. Then the probability that all components will fail is much smaller (usually at least one order of magnitude) than the probability of failure of each component.

3.7 Random variables

In order to be able to handle the events in a sample space mathematically, it is useful to attach a numerical value to each elementary event. (Recall that elementary events are events that contain just one point.) The set of numerical values of the elementary events is called a *random variable*, usually abbreviated as r.v. Conventionally, random variables are represented by capital letters, e.g. X.

When the elementary events are themselves numbers, there is a natural way of doing this. But it is sometimes useful to attach numbers to qualitative events. For example, suppose the sample space considered is that of the sex of an occupant and consists of two possible outcomes, namely {*male, female*}. We could assign to the event "male" the number 0 and to the event "female" the number 1. The advantage of doing this is, for example, that if we consider n occupants and attach to occupant i the random variable X_i, then the number of females among the n occupants can be represented as $X_1 + X_2 + \cdots + X_n$ (or equivalently by $\sum X_i$).

3.7.1 Types of random variables

In this work we shall consider two types of random variables: discrete and continuous.

A random variable is said to be *discrete* if its range is a discrete (finite or countable) set of real numbers. Usually, the values are $\{0, 1, 2, \ldots\}$. It is said to be *continuous* if its range is an interval of the real line (finite or infinite).

Examples

1. *Discrete random variables*: Number of rooms in a building, number of fires in a city in 1 year, number of absent firemen in a fire brigade on a specific date, number of defective sprinklers in a factory.
2. *Continuous random variables*: Weight of fuel in a room, maximum temperature reached by a fire, ultimate strength of a beam at some high temperature, speed of spread of smoke across some opening at a particular time.

3.7.2 Discrete random variables

The probability function

It is not necessary for different elementary events in a sample space to have different values of the random variable corresponding to them. Whether or not this is the case, we attach to each value x of a discrete random variable X the sum of the probabilities of all the elementary events to which is attached the considered value. This sum is called the *probability function* of the random variable and is denoted by $P(X = x)$.

Example   Consider a room in which there are three occupants, denoted by P, Q and R. We are interested in the number of children among them, denoted by X. Denote children by C and adults by A. Table 3.1 lists all elementary events, their probabilities and the value of X for each.

From this table, we can obtain the values of the probability function of X, as follows:

1. $P(X = 0) = 0.336$,
2. $P(X = 1) = 0.084 + 0.144 + 0.224 = 0.452$,
3. $P(X = 2) = 0.036 + 0.056 + 0.096 = 0.188$,
4. $P(X = 3) = 0.024$.

It is usual to present the probability function in tabular form, as in Table 3.2.

Table 3.1 Elementary events, probabilities and value of random variable

P	Q	R	Probability	X
A	A	A	0.336	0
A	A	C	0.084	1
A	C	A	0.144	1
A	C	C	0.036	2
C	A	A	0.224	1
C	A	C	0.056	2
C	C	A	0.096	2
C	C	C	0.024	3

Table 3.2 Probability function of X

x	0	1	2	3
$P(X=x)$	0.336	0.452	0.188	0.024

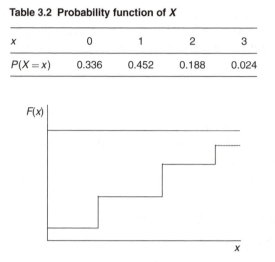

Figure 3.4 Distribution function of a discrete variable.

Properties of the probability function   The probability function clearly has the following two properties:

1. $0 \le P(X=x) \le 1$.
2. $\sum_x P(X=x) = 1$.

The distribution function

The *distribution function* of X, also called the *cumulative probability*, and denoted by $F(x)$, is defined by

$$F(x) = P(X \le x) \tag{3.19}$$

$$= \sum_{t \le x} P(X = t). \tag{3.20}$$

In the example of the previous subsection, we have

$F(0) = P(X \le 0) = P(X = 0) = 0.336,$

$F(1) = P(X \le 1) = P(X = 0) + P(X = 1) = 0.788,$

$F(2) = P(X \le 2) = P(X = 0) + P(X = 1) + P(X = 2) = 0.976,$

$F(3) = P(X \le 3) = P(X = 0) + P(X = 1) + P(X = 2) + P(X = 3) = 1.$

However, unlike the probability function, the distribution function is defined for all real values of x. For example, we have $F(1.5) = P(X \le 1.5) = 0.788$. And, of course, $F(-1) = 0$ and $F(4) = 1$. In fact, $F(x)$ is a non-decreasing function of x such that $F(-\infty) = 0$ and $F(+\infty) = 1$. In the case of a discrete random variable, it is a step function (see Figure 3.4).

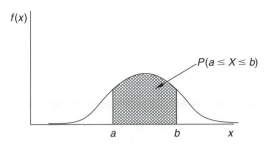

Figure 3.5 Probability density function.

3.7.3 Continuous random variables

Continuous random variables are represented mathematically by *probability density functions*. A function $f(x)$ is called the *probability density function* (abbreviated p.d.f.) of the random variable X if for all $a \leq b$:

$$P(a \leq X \leq b) = \int_a^b f(x)\, dx. \tag{3.21}$$

In other words, the area under the curve of $f(x)$ between the ordinates a and b represents the probability that $a \leq X \leq b$. This is illustrated in Figure 3.5.

Clearly, $f(x)$ must satisfy the following two conditions:

1. $f(x) \geq 0$.
2. $\int_{-\infty}^{+\infty} f(x)\, dx = 1$.

It also follows from the definition that the probability of X assuming a particular value is zero, since

$$P(X = b) = P(b \leq X \leq b) \tag{3.22}$$

$$= \int_b^b f(x)\, dx \tag{3.23}$$

$$= 0. \tag{3.24}$$

As a corollary, we have, for a continuous random variable,

$$P(a \leq X \leq b) = P(a < X \leq b) = P(a \leq X < b) = P(a < X < b).$$

Example   Let X be the velocity, in m/s, at which a flame spreads along a slab of fuel of unit width, and let the p.d.f. of X be

$$f(x) = 6x(1 - x) \quad \text{for } 0 \leq x \leq 1, \tag{3.25}$$

$$= 0 \qquad\qquad \text{otherwise.} \tag{3.26}$$

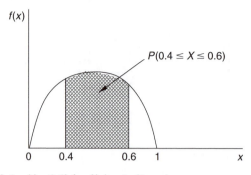

Figure 3.6 $P(0.4 \le X \le 0.6)$ for $f(x) = 6x(1 - x)$.

Clearly $f(x)$ is a valid p.d.f., since

1. $f(x) \ge 0$ for all x.

2.
$$\int_{-\infty}^{+\infty} f(x)\, dx = \int_{-\infty}^{0} 0\, dx + \int_{0}^{1} 6x(1 - x)\, dx + \int_{1}^{+\infty} 0\, dx \qquad (3.27)$$

$$= \int_{0}^{1} (6x - 6x^2)\, dx \qquad (3.28)$$

$$= [3x^2 - 2x^3]_0^1 = 3 - 2 = 1. \qquad (3.29)$$

Let us find $P(0.4 \le X \le 0.6)$ (Figure 3.6).
    We have

$$P(0.4 \le X \le 0.6) = \int_{0.4}^{0.6} 6x(1 - x)\, dx \qquad (3.30)$$

$$= [3x^2 - 2x^3]_{0.4}^{0.6} \qquad (3.31)$$

$$= 0.296. \qquad (3.32)$$

Distribution function of a continuous variable

The distribution function $F(x)$ of a continuous random variable X with p.d.f. $f(x)$ is given by

$$F(x) = P(X \le x) = \int_{-\infty}^{x} f(t)\, dt \qquad (3.33)$$

It is illustrated in Figure 3.7.

Remark   Knowledge of the distribution function $F(x)$ of a random variable X enables probabilities such as $P(a < X < b)$ to readily evaluated, since, obviously, $P(a < X < b) = F(b) - F(a)$.

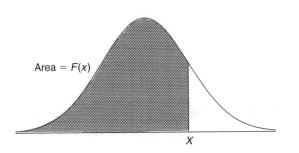

Figure 3.7 Distribution function $F(x)$ of X.

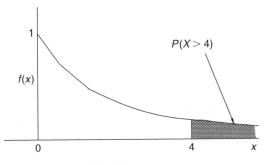

Figure 3.8 Exponential density function.

Example   Let X be the time in years until a new smoke alarm battery requires replacement, and let its p.d.f. be $f(x) = e^{-x}$ for $x \geq 0$ and $f(x) = 0$ for $x < 0$ (Figure 3.8):

1. Clearly $f(x) \geq 0$ and

$$\int_{-\infty}^{\infty} f(x)\, dx = \int_{0}^{\infty} e^{-x}\, dx = 1 \qquad (3.34)$$

so $f(x)$ is a valid p.d.f.

2. $\qquad P(X > 4) = \int_{4}^{\infty} e^{-x}\, dx = [-e^{-x}]_{4}^{\infty} = e^{-4} = 0.0183. \qquad (3.35)$

3. For $x \geq 0$

$$F(x) = \int_{0}^{x} e^{-t}\, dt = [-e^{-t}]_{t}^{x} = e^{-4} = 1 - e^{-x} \qquad (3.36)$$

and for $x < 0$, $F(x) = 0$.

4. $\qquad\qquad P(X < 1) = F(1) = 1 - e^{-1} = 0.6321. \qquad (3.37)$

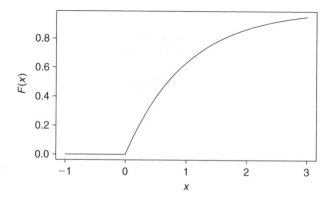

Figure 3.9 Exponential distribution function.

3.7.4 Quantiles

Let the random variable X have the distribution $F(x)$. Then the *quantile q* corresponding to the probability p is defined as the solution of the equation $F(q) = p$. In other words, q is the value of X such that $P(X \leq q) = p$. The *median* is the quantile corresponding to $p = 1/2$. Other frequently used quantiles are the lower and upper *quartiles*, corresponding to $p = 1/4$ and $p = 3/4$, respectively (Figure 3.9).

Example   Let $F(x) = 1 - e^{-x}$ (corresponding to $f(x) = e^{-x}$ for $x > 0$ and 0 otherwise). Then the 95% quantile is obtained by solving the equation

$$1 - e^{-q} = 0.95, \qquad (3.38)$$

from which $q = -\log(0.05) = 2.996$.

3.8 Mathematical expectation

Mathematical expectation is an averaging process on a random variable, with the weights being the probabilities of the various outcomes occurring. The expected value of a random variable X is denoted by $E(X)$. It is also known simply as the *expectation* or *mean* of X. The concept of expectation is similar to the concept of centre of gravity in mechanics.

Examples

1. Consider the discrete random variable X introduced in Section 3.7.2, which represents the number of children among three occupants of a room. The probability function of X is given in Table 3.2.
   The expected value of X is given by

$$E(X) = (0 \times 0.336) + (1 \times 0.452) + (2 \times 0.188) + (3 \times 0.024) \quad (3.39)$$

$$= 0.9. \qquad (3.40)$$

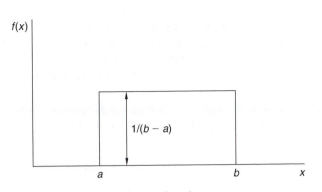

Figure 3.10 Uniform probability density function.

Thus the average ("expected") number of children in the room is 0.9.

In general, if X is a discrete random variable with probability function $f(x)$ then the expected value $E(X)$ is given by

$$E(X) = \sum_x x f(x). \tag{3.41}$$

If, on the other hand, X is a continuous random variable with probability density function $f(x)$:

$$E(X) = \int_{-\infty}^{\infty} x f(x) \, dx. \tag{3.42}$$

2. Consider the uniform probability density function over (a, b) (Figure 3.10):

$$f(x) = \begin{cases} 1/(b-a) & \text{if } a < x < b, \\ 0 & \text{otherwise.} \end{cases} \tag{3.43}$$

We have

$$E(X) = \int_{-\infty}^{\infty} x f(x) \, dx \tag{3.44}$$

$$= \frac{1}{(b-a)} \int_a^b x \, dx \tag{3.45}$$

$$= \frac{1}{(b-a)} \left[\frac{1}{2} x^2 \right]_a^b \tag{3.46}$$

$$= \frac{1}{(b-a)} \frac{1}{2} (b^2 - a^2) \tag{3.47}$$

$$= \frac{1}{(b-a)} \frac{1}{2} (b+a)(b-a) \tag{3.48}$$

$$= \frac{(b+a)}{2}. \tag{3.49}$$

i.e. the mean of X is at the midpoint of (a, b), as would be expected from considerations of symmetry. In general, the mean of a random variable whose p.d.f. is symmetric about some point x_0 is at x_0.

The above definition readily generalizes to the expectation of any function $g(x)$ of X.

$$E[g(X)] = \sum_x g(x) f(x) \quad \text{if } X \text{ is discrete,} \tag{3.50}$$

$$= \int_{-\infty}^{\infty} g(x) f(x) \, dx \quad \text{if } X \text{ is continuous.} \tag{3.51}$$

For the above example, take $g(X) = X^2$. Then

$$E(X^2) = \int_a^b x^2 \frac{1}{(b-a)} \, dx \tag{3.52}$$

$$= \frac{b^2 + ab + a^2}{3}. \tag{3.53}$$

3.8.1 Laws of expectation

1. If b is a constant, then $E(b) = b$.
2. If a and b are constants, then

$$E(aX + b) = aE(X) + b. \tag{3.54}$$

In other words, the expectation operator is *linear*.

More generally

$$E[g(X) + h(X)] = E[g(X)] + E[h(X)]. \tag{3.55}$$

3.8.2 Moments

Definition 3.8.1. *The expectation of the kth power of a random variable (for integer k) is called the kth moment of the random variable, and is denoted by* μ'_k, *i.e.:*

$$\mu'_k = E(X^k) \tag{3.56}$$

$$= \sum_x^k f(x) \quad \text{for } X \text{ discrete,} \tag{3.57}$$

$$= \int_{-\infty}^{\infty} x^k f(x) \, dx \quad \text{for } X \text{ continuous.} \tag{3.58}$$

Note that:

1. $\mu'_0 = E(X^0) = E(1) = 1$.
2. $\mu'_1 = E(X^1) = E(X)$ is called, as mentioned at the beginning of Section 3.8, the *mean* of X.

The mean μ'_1 is usually denoted simply by μ.

3.8.3 Moments about the mean (central moments)

It is useful to consider moments about the mean of a random variable. The kth moment about the mean of the random variable X is

$$\mu_k = E[(X - \mu)^k] \tag{3.59}$$

$$= \sum_x (x - \mu)^k P(X = x) \quad \text{for } X \text{ discrete,} \tag{3.60}$$

$$= \int_{-\infty}^{\infty} (x - \mu)^k f(x)\, dx \quad \text{for } X \text{ continuous.} \tag{3.61}$$

For $k = 1$ we have, using the laws of expectation given in Section 3.8.1,

$$\mu_1 = E[X - \mu) \tag{3.62}$$

$$= E(X) - \mu \tag{3.63}$$

$$= 0. \tag{3.64}$$

3.8.4 Variance

The second moment about the mean is of particular interest, as it is a measure of variability or dispersion of the values of the random variable. It is commonly denoted by Var(X) or σ^2.

Theorem 3.8.1. $\sigma^2 = E(X^2) - [E(X)]^2$.

Proof

$$\sigma^2 = E[(X - \mu)^2] \tag{3.65}$$

$$= E(X^2 - 2\mu X + \mu^2) \tag{3.66}$$

$$= E(X^2) - 2\mu E(X) + \mu^2 \tag{3.67}$$

$$= E(X^2) - 2\mu^2 + \mu^2 \tag{3.68}$$

$$= E(X^2) - \mu^2. \tag{3.69}$$

□

The positive square root of σ^2, denoted by σ, is called the standard deviation of X. Note that σ has the same units as X.

3.8.5 Variance of a linear function of a random variable

Consider the linear function of the random variable X: $aX + b$, where a and b are constants.

Theorem 3.8.2. $\mathrm{Var}(aX + b) = a^2\mathrm{Var}(X)$.

Proof

$$\mathrm{Var}(aX + b) = E[\{aX + b - E(aX + b)\}^2] \tag{3.70}$$

$$= E[\{aX - aE(X)\}^2] \tag{3.71}$$

$$= a^2 E[\{X - E(X)\}^2] \tag{3.72}$$

$$= a^2\mathrm{Var}(X). \tag{3.73}$$

□

Notes

1. Adding a constant to a random variable does not change the variance.
2. Multiplying a random variable by a constant a multiplies the variance by a^2.
3. The *coefficient of variation* of a random variable is equal to the standard deviation divided by the mean. It is important to note that it is a dimensionless quantity.

3.9 Jointly distributed random variables

Suppose we define *two* random variables X and Y on the sample space S. Thus each point in S has a value for X and a value for Y. Then X and Y are said to be jointly distributed.

If X and Y are both discrete random variables they have a *joint probability function* $f(x, y) = P(X = x, Y = y)$ with the properties:

1. $f(x, y) \geq 0$ for all x, y.
2. $\sum_{\text{all } x} \sum_{\text{all } y} f(x, y) = 1$.
3. $P(X \leq x, Y \leq y) = F(x, y) = \sum_{s \leq x} \sum_{t \leq y} f(s, t)$.

If X and Y are both continuous random variables they have a *joint probability density function* $f(x, y)$ with the properties:

1. $f(x, y) \geq 0$ for all x, y.
2. $\int_{-\infty}^{\infty} \int_{-\infty}^{\infty} f(x, y) \, dx \, dy = 1$.
3. $P(X \leq x, Y \leq y) = F(x, y) = \int_{-\infty}^{x} \int_{-\infty}^{y} f(s, t) \, ds \, dt$.

It is often the case when X and Y are discrete that they only take each a small number of values. In that case, it is usually convenient to represent the joint distribution by a table.

Table 3.3 Joint probability function of fires and fire brigade calls

y	x		
	0	1	2
0	0.05	0.05	0.1
1	0.05	0.1	0.35
2	0	0.2	0.1

Examples

1. Two discrete random variables.

   Let X be the number of fires in 1 day in a certain district and let Y be the number of fire brigade calls. Suppose X and Y take only the values 0, 1 and 2. The joint probability function is given by Table 3.3.

2. Two continuous random variables.

   In the gaseous combustion product of an experimental fire let X be the proportion of carbon monoxide and Y the proportion of carbon dioxide. Let the joint density function of X and Y be

$$f(x,y) = \begin{cases} \frac{2}{5}(2x + 3y) & \text{if } 0 \leq x \leq 1,\ 0 \leq y \leq 1, \\ 0 & \text{elsewhere.} \end{cases} \tag{3.74}$$

It is easy to verify that the total probability is 1. Indeed

$$\int_{-\infty}^{\infty} \int_{-\infty}^{\infty} f(x,y)\, dx\, dy = \int_0^1 \int_0^1 \frac{2}{5}(2x + 3y)\, dx\, dy \tag{3.75}$$

$$= 1. \tag{3.76}$$

3.9.1 Marginal distributions

The probability density function (or probability function) of each of the random variables alone can be obtained from the joint probability density function (or probability function) by integrating (or summing) over the other random variable. The resulting probability density function (or probability function) is called the *marginal* probability density function (or probability function).

Given $f(x,y)$, the marginal probability density function (or probability function) of X is given by $g(x)$, where

$$g(x) = \sum_{\text{all } y} f(x,y) \quad \text{for } X, Y \text{ discrete} \tag{3.77}$$

and

$$g(x) = \int_{-\infty}^{\infty} f(x,y)\, dy \quad \text{for } X, Y \text{ continuous.} \tag{3.78}$$

Table 3.4 Marginal probability functions of fires and fire brigade calls

y	x			
	0	1	2	
0	0.05	0.05	0.1	0.2
1	0.05	0.1	0.35	0.5
2	0	0.2	0.1	0.3
	0.1	0.35	0.55	

Examples

1. Consider the joint distribution of fires and fire brigade calls given in the last section. The marginal probability functions are obtained as the row and column sums of the joint probability function (see Table 3.4).
2. Consider the joint p.d.f. of carbon monoxide and carbon dioxide given in example 2 of the previous section. Let $g(x)$ be the marginal p.d.f. of X. Then

$$g(x) = \int_{-\infty}^{\infty} f(x,y)\,dy = \int_{0}^{1} \frac{2}{5}(2x+3y)\,dy \qquad (3.79)$$

$$= \frac{4x+3}{5}. \qquad (3.80)$$

Similarly, let $h(y)$ be the marginal p.d.f. of Y. Then

$$h(y) = \int_{-\infty}^{\infty} f(x,y)\,dx = \int_{0}^{1} \frac{2}{5}(2x+3y)\,dx \qquad (3.81)$$

$$= \frac{2(1+3y)}{5}. \qquad (3.82)$$

3.9.2 Conditional distributions

The conditional distribution of X, given that $Y = y$, is defined by

$$f(x|y) = \frac{f(x,y)}{h(y)}, \qquad (3.83)$$

where $h(y)$ is the marginal distribution of Y. (Recall $P(A|B) = P(A \cap B)/P(B)$.)

Examples

1. Discrete random variable.

   Consider again the case of the joint distribution of number of fires and fire brigade calls (Table 3.4). The probability that there were no fire brigade calls, given that there was one fire, is given by

$$\frac{f(1,0)}{g(1)} = \frac{0.05}{0.35} = 0.143. \qquad (3.84)$$

2. Continuous random variable.

In example 2 of the previous section, the p.d.f. of the carbon monoxide proportion, given that the proportion of carbon dioxide is $Y = y$, is given by

$$f(x|y) = \frac{f(x, y)}{h(y)} \tag{3.85}$$

$$= \frac{2x + 3y}{1 + 3y}. \tag{3.86}$$

Thus if the carbon dioxide proportion Y is 0.5 the value of the conditional p.d.f. of the carbon monoxide at $X = 0.3$ is $(2 \times 0.3 + 3 \times 0.5)/(1 + 3 \times 0.5) = 0.84$.

3.10 Independence

Recall the definition of two independent events A and B:

$$P(A|B) = P(A). \tag{3.87}$$

(which is equivalent to $P(A \cap B) = P(A)P(B)$)

We extend this to independence of random variables as follows: X and Y are independent if

$$f(x|y) = g(x) \tag{3.88}$$

for every relevant value of X and Y, where $g(x)$ is the marginal distribution of x, i.e., knowing that $Y = y$ has no effect on the distribution of X.

From the definition of conditional probability it follows that

$$f(x, y) = g(x)h(y) \tag{3.89}$$

where $h(y)$ is the marginal distribution of Y. In words: for discrete random variables the joint probability function is the product of the marginal probability functions. For continuous random variables the joint probability density function is the product of the marginal p.d.f.s.

Examples

1. Discrete random variables.

Consider the joint distribution of fires and fire brigade calls given in Table 3.4. We have $P(X = 1, Y = 1) = 0.1$. On the other hand, $P(X = 1) = 0.35$ and $P(Y = 1) = 0.5$. Thus $P(X = 1, Y = 1) \neq P(X = 1)$ $P(Y = 1)$ and X and Y are *not* independent.

2. Continuous random variables.

Consider example 2 of Section 3.9. In the square $0 \leq x \leq 1, 0 \leq y \leq 1$ the joint p.d.f. of X (the proportion of carbon monoxide) and Y (the proportion of carbon dioxide) is $\frac{2}{5}(2x + 3y)$. In Section 3.9.1 it was shown that the marginal distributions of X and Y are $(4x+3)/5$ and $(2+6y)/5$, respectively. Since the joint probability density function is clearly not the product of the marginal p.d.f.s., X and Y are *not* independent.

3. Independent continuous random variables.

Let X and Y be the lifetime in years of two smoke alarm batteries made by two different manufacturers. Let the joint distribution of X and Y be

$$f(x, y) = \begin{cases} 0.2e^{-0.5x-0.4y} & \text{for } 0 \le x \le \infty, 0 \le y \le \infty, \\ 0 & \text{elsewhere.} \end{cases} \tag{3.90}$$

The marginal p.d.f. of X is given by

$$g(x) = \int_0^\infty 0.2e^{-0.5x-0.4y}\, dy \tag{3.91}$$

$$= 0.5e^{-0.5x} \quad \text{for } 0 \le x \le \infty, \tag{3.92}$$

while the marginal p.d.f. of Y is given by

$$h(y) = \int_0^\infty 0.2e^{-0.5x-0.4y}\, dx \tag{3.93}$$

$$= 0.4e^{-0.4y} \quad \text{for } 0 \le x \le \infty. \tag{3.94}$$

Here we have $f(x, y) = g(x)h(y)$, so that X and Y are independent, as would be reasonably expected, since knowing the exact lifetime of battery 1 does not give any information about the lifetime distribution of battery 2.

3.10.1 Bivariate moments

Let X and Y be jointly distributed random variables with joint p.f. (or p.d.f.) $f(x, y)$ and means μ_X, μ_Y, respectively. Then we define

$$E(XY) = \sum_x \sum_y xy f(x, y) \quad \text{for } X \text{ and } Y \text{ discrete,} \tag{3.95}$$

$$= \int_{-\infty}^\infty \int_{-\infty}^\infty xy f(x, y)\, dx\, dy \quad \text{for } X \text{ and } Y \text{ continuous.} \tag{3.96}$$

Covariance

The joint first moment of X and Y about their respective means, $E[(X - \mu_X)(Y - \mu_Y)]$, is called the *covariance* of X and Y, is denoted by $\text{Cov}(X, Y)$ and is a measure of their association.

Recall that for the variance of a random variable $\sigma^2 = E(X^2) - \mu^2$. Similarly, for the covariance:

$$\text{Cov}(X, Y) = E[(X - \mu_X)(Y - \mu_Y)] \tag{3.97}$$

$$= E[XY - X\mu_Y - Y\mu_X + \mu_X\mu_Y] \tag{3.98}$$

$$= E(XY) - E(X)\mu_Y - E(Y)\mu_X + \mu_X\mu_Y \tag{3.99}$$

$$= E(XY) - \mu_X\mu_Y. \tag{3.100}$$

Positive covariance means that X and Y vary together, i.e. high X implies high Y and conversely low X implies low Y.

Similarly, negative covariance means that X and Y vary oppositely, i.e. high X implies low Y and conversely low X implies high Y.

Examples

- *Positive covariance*
  - Amount of fuel in a room and maximum temperature reached in a fire.
  - Fire severity and amount of damage caused.
- *Negative covariance*
  - Width of stairways in a multistorey building and number of casualties in a fire.
  - Number of automatic sprinklers in a warehouse and fire duration.

Theorem 3.10.1.   *If X and Y are independent $E(XY) = E(X)E(Y)$.*

Proof   The proof will be given for continuous random variables. A similar proof holds for discrete variables:

$$E(X, Y) = \int_{-\infty}^{\infty} \int_{-\infty}^{\infty} x\,yf(x, y)\,\mathrm{d}x\,\mathrm{d}y \tag{3.101}$$

$$= \int_{-\infty}^{\infty} \int_{-\infty}^{\infty} xy\,g(x)\,h(y)\,\mathrm{d}x\,\mathrm{d}y \tag{3.102}$$

$$= \left[\int_{-\infty}^{\infty} x\,g(x)\,\mathrm{d}x\right]\left[\int_{-\infty}^{\infty} y\,h(y)\,\mathrm{d}y\right] \tag{3.103}$$

$$= E(X)E(Y). \tag{3.104}$$

$\square$

Corollary 1.   *It follows that if X and Y are independent*

$$\text{Cov}(X, Y) = E[(X - \mu_X)(Y - \mu_Y)] \tag{3.105}$$

$$= E(X - \mu_X)E(Y - \mu_Y) \tag{3.106}$$

$$= 0. \tag{3.107}$$

Thus X and Y independent implies that $\text{Cov}(X, Y) = 0$. *The converse is not necessarily true but* $\text{Cov}(X, Y) = 0$ *is usually taken as a strong indication that X and Y are independent.*

Examples   Find the covariance of the carbon monoxide and carbon dioxide proportions in example 2 of Section 3.9.
   We have

$$E(XY) = \int_0^1 \int_0^1 xy\frac{2}{5}(2x + 3y)\,dx\,dy \tag{3.108}$$

$$= \frac{1}{3}. \tag{3.109}$$

Furthermore

$$E(X) = \int_0^1 x\frac{2}{5}(2x + 3y)\,dx\,dy \tag{3.110}$$

$$= \frac{17}{30} \tag{3.111}$$

and

$$E(Y) = \int_0^1 y\frac{2}{5}(2x + 3y)\,dx\,dy \tag{3.112}$$

$$= \frac{3}{5} \tag{3.113}$$

so that finally

$$\text{Cov}(X, Y) = E(XY) - E(X)E(Y) \tag{3.114}$$

$$= \frac{1}{3} - \frac{17}{30}\frac{3}{5} \tag{3.115}$$

$$= -\frac{1}{150}. \tag{3.116}$$

As intuitively expected, the covariance is negative.

Rules for bivariate second moments

Theorem 3.10.2.

$$\text{Var}(aX + bY) = a^2\text{Var}(X) + b^2\text{Var}(Y) + 2ab\text{Cov}(X, Y). \tag{3.117}$$

Proof

$$\mathrm{Var}(aX + bY) = E([aX + bY - E\{aX + bY\}]^2)$$
$$= E([aX + bY - aE(X) - bE(Y)]^2)$$
$$= E([a(X - \mu_X) + b(Y - \mu_Y)]^2)$$
$$= a^2 E[(X - \mu_X)^2] + b^2 E[(Y - \mu_Y)^2]$$
$$+ 2ab E[(X - \mu_X)(Y - \mu_Y)]$$
$$= a^2 \mathrm{Var}(X) + b^2 \mathrm{Var}(Y) + 2ab\mathrm{Cov}(X, Y). \qquad \square$$

Corollary 2.   *If X and Y are independent, then $\mathrm{Cov}(X, Y) = 0$ and hence*

$$\mathrm{Var}(aX + bY) = a^2\mathrm{Var}(X) + b^2\mathrm{Var}(Y). \qquad (3.118)$$

Corollary 3.   *More generally, if $X_1, X_2, \ldots, X_n$ are independent random variables, with $E(X_i) = \mu_i$ and $\mathrm{Var}(X_i) = \sigma_i^2$, then*

$$E\left(\sum_{i=1}^{n} X_i\right) = \sum_{i=1}^{n} E(X_i)$$
$$= \sum_{i=1}^{n} \mu_i$$

and

$$\mathrm{Var}\left(\sum_{i=1}^{n} X_i\right) = \sum_{i=1}^{n} \mathrm{Var}(X_i) \qquad (3.119)$$
$$= \sum_{i=1}^{n} \sigma_i^2. \qquad (3.120)$$

Correlation

It is often convenient to measure the dependence between two random variables by means of a dimensionless parameter, the *correlation coefficient*.

Let X and Y have standard deviations σ_X and σ_Y, respectively. Their correlation coefficient ρ_{XY} is defined by

$$\rho_{XY} = \frac{\mathrm{Cov}(X, Y)}{\sigma_X \sigma_Y}. \qquad (3.121)$$

The correlation coefficient ρ has the following properties:

1. It is independent of scale and origin.
2. $-1 \leq \rho \leq 1$.

3. If X and Y are independent $\rho_{XY} = 0$.
4. If $|\rho| = 1$ then we must have $Y = aX + b$ for some a and b, i.e. Y is fully dependent on X.

3.10.2 Matrix formulation

We can think of a set of n random variables $(X_1, X_2, \ldots, X_n)$ as a random column vector of length n and denote it by $\mathbf{X}$. We shall denote the mean of the vector $\mathbf{X}$ by $\boldsymbol{\mu} = (\mu_1, \mu_2, \ldots, \mu_n)$.

Let the transpose of the matrix $\mathbf{A}$ be denoted by $\mathbf{A}^T$. Consider the $n \times n$ matrix

$$\boldsymbol{\Sigma} = E[(\mathbf{X} - \boldsymbol{\mu})(\mathbf{X} - \boldsymbol{\mu})^T]. \tag{3.122}$$

It is called the *covariance matrix* of the vector $\mathbf{X}$. We note that the (i, j)th element of $\boldsymbol{\Sigma}$ is $E[(X_i - \mu_i)(X_j - \mu_j)]$, i.e. the covariance of X_i and X_j. The diagonal elements are simply the variances of the X_i.

If we replace the X_i by *reduced variables*

$$Z_i = \frac{X_i - \mu_i}{\sigma_i} \tag{3.123}$$

then the covariance matrix of the Z_i, denoted by $\mathbf{C}$, has as its (i, j)th element the *correlation* ρ_{ij} of X_i and X_j. The diagonal elements are all equal to 1.

Suppose now that we have a random vector $\mathbf{Y} = \mathbf{A}\mathbf{X}$ where $\mathbf{A}$ is some matrix. The covariance matrix of $\mathbf{Y}$, $\boldsymbol{\Sigma}_{\mathbf{Y}}$ will be given by

$$\boldsymbol{\Sigma}_{\mathbf{Y}} = E[(\mathbf{Y} - \boldsymbol{\mu}_{\mathbf{Y}})(\mathbf{Y} - \boldsymbol{\mu}_{\mathbf{Y}})^T]$$

$$= \mathbf{A}E[(\mathbf{X} - \boldsymbol{\mu}_{\mathbf{X}})(\mathbf{X} - \boldsymbol{\mu}_{\mathbf{X}})^T]\mathbf{A}^T$$

$$= \mathbf{A}\boldsymbol{\Sigma}_{\mathbf{X}}\mathbf{A}^T. \tag{3.124}$$

Equation (3.124) is the generalization of Theorem 3.10.2.

3.10.3 Sampling

When an experiment is repeated n times, say, and the value of some random variable is measured for each repetition, we obtain a *random sample* of values of the random variable. This sample is conceptually thought of as a collection of n independent, identically distributed random variables $(X_1, X_2, \ldots, X_n)$.

Random samples from discrete distributions are usually represented by a bar chart, with a bar for each value of the random variable. Since probability is defined as relative frequency in a large number of repetitions, the relative size of the bars can be made to approximate the relative probability of the values of the random variable as nearly as required by taking the sample size large enough.

For continuous distributions it is usual to represent the sample by a special type of bar chart called a *histogram*. The range of the sample is divided into an appropriate number of equal intervals, and the bars represent the number of sample elements that fall in each interval. The relative number approximates the probability that the random variable falls in the interval. For comparison, the theoretical density function, suitably scaled, is often superimposed on the histogram.

A number of illustrations of histograms, where the value of $n\delta f(x)$ is plotted (n is the size of the sample and δ, the width of the interval) will be given in the next section.

3.10.4 Quantiles of a random sample

The quantiles of a distribution are defined in Section 3.7.4.

To calculate the quantiles of a random sample, it is first rearranged in ascending order, renaming the elements of the sample $X_{[i]}$. Thus

$$X_{[1]} \le X_{[2]} \le \cdots \le X_{[n]}. \tag{3.125}$$

Then $q = X_{[i]}$ is the quantile corresponding to $p = i/n$. This defines n quantiles.

3.10.5 Quantile–quantile plot

A quantile–quantile plot is a device to check whether a random sample can be thought of as deriving from a given probability distribution. It is usual to compare the sample with a random variable of the form $a + bU$ where U is a standard random variable (e.g. standard normal) and a and $b > 0$ are unknown constants. The sample quantiles are plotted on the y-axis and the corresponding quantiles of U are plotted on the x-axis. If the considered distribution is suitable, the plot should fall on the line $y = a + bx$. An example of the use of a quantile–quantile plot is given in Section 8.3.

3.11 Some probability distributions

3.11.1 Discrete probability distributions

The Bernoulli variable

Consider a random experiment in which there are only two possible outcomes. Call these outcomes *success* and *failure*. Such an experiment is called a *Bernoulli trial*. Let X be a random variable which takes the value 1 when the outcome is a success and 0 when the outcome is a failure. It is called a *Bernoulli variable*.

Examples

1. Whether a particular occupant will die in a fire.
2. Whether or not flashover will occur in a compartment fire.
3. Whether a particular sprinkler will operate or not.

Let the probability of success be denoted by p and the probability of failure by $q = 1 - p$.

Clearly we have $E(X) = 1 \times p + 0 \times (1 - p) = p$. Moreover $E(X^2) = 1^2 \times p + 0^2 \times (1 - p) = p$. It follows that

$$\text{Var}(X) = E(X^2) - [E(X)^2] \qquad (3.126)$$

$$= p - p^2 \qquad (3.127)$$

$$= p(1 - p). \qquad (3.128)$$

The binomial distribution

Consider now ν independent repetitions of the experiment ("ν independent trials") and define the random variable X as the number of successes in ν independent trials.

Clearly, X can assume the values $0, 1, \ldots, \nu$. The probability of x successes (and therefore $\nu - x$ failures) is given by the formula:

$$P(X = x) = \begin{cases} \binom{\nu}{x} p^x (1 - p)^{\nu - x} & \text{for } x = 0, 1, \ldots, \nu; \\ 0 & \text{otherwise}; \end{cases} \qquad (3.129)$$

where $\binom{\nu}{x}$, the *binomial coefficient* is given by

$$\binom{\nu}{x} = \frac{\nu!}{x!(\nu - x)!} \qquad (3.130)$$

and $x! = 1.2.3 \ldots (x - 1).x$. Note that $0! = 1! = 1$ and therefore $\binom{\nu}{0} = \binom{\nu}{\nu} = 1$.

This is called the *binomial distribution* with parameters ν and p, and is denoted as

$$X \sim B(\nu, p), \qquad (3.131)$$

where $\sim$ means "is distributed as".

Examples   There are six occupants in an apartment building. Suppose that the probability of dying for each occupant is 0.2 and that the risks of death for different occupants are independent of each other. We require the probabilities that 0, 1, 2, 3, 4, 5 or 6 occupants die.

If X denotes the number of occupants dying, then $X \sim B(6, 0.2)$, and

$$P(X = x) = \binom{6}{x} 0.2^x 0.8^{6-x} \quad \text{for } x = 0, 1, 2, 3, 4, 5, 6. \qquad (3.132)$$

The probability function is given by Table 3.5.

Figure 3.11 shows a bar plot of the binomial distribution.

Table 3.5 Probability function of X

x	0	1	2	3	4	5	6
$P(X = x)$	0.2621	0.3932	0.2458	0.0819	0.0154	0.0015	0.000064

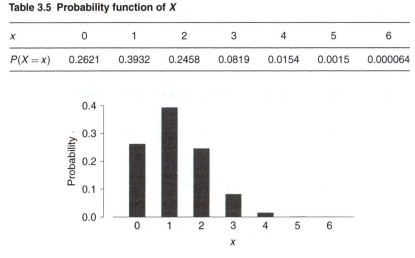

Figure 3.11 Probability function of binomial distribution.

Mean and variance of a binomial random variable The mean of the binomial distribution is given by

$$E(X) = vp, \tag{3.133}$$

and the variance by

$$\text{Var}(X) = vp(1 - p), \tag{3.134}$$

so that the standard deviation is $\sqrt{vp(1-p)}$.

This can be easily seen if we note that we can write $X = X_1 + X_2 + \cdots + X_v$ where the X_i are independent Bernoulli variables, each with probability of success p. The formulae for the mean and variance of X then follow immediately from formulae (3.119) and (3.120).

For details see for example Devore [20, pp. 117–119].

The Poisson distribution

The Poisson random variable counts the number of occurrences of rare events in a time interval, a region of space (one-, two- or three-dimensional) or a collection of numerous objects, all of fixed size.

Examples

1. The number of fires causing damage of more than $100,000 in a suburb per year.
2. The number of fire stations per square kilometre in a large metropolis.
3. The number of fire extinguishers sold by a retail shop per year.
4. The number of children per family in a large apartment block.
5. The number of flaws in a steel girder.

Assumptions underlying the Poisson distribution   The assumptions will be framed in terms of time intervals, but similar assumptions can be framed for space regions or collections of objects:

1. The number of occurrences of events in one time interval is independent of the number occurring in any other non-overlapping interval.
2. The probability of occurrence of an event in a small interval is proportional to the length of that interval.
3. The probability of more than one event occurring in a small time interval is negligible.

Probability function   The probability function of a Poisson random variable X is given by

$$P(X = x) = \begin{cases} e^{-\lambda}\lambda^x/x! & \text{for } x = 0, 1, 2, \ldots, \\ 0 & \text{otherwise.} \end{cases} \tag{3.135}$$

The parameter λ is the average occurrence rate for the given interval size. We write $X \sim P(X)$. Figure 3.12 shows a bar plot of the distribution for $\lambda = 1$, Figure 3.13 for $\lambda = 5$ and Figure 3.14 for $\lambda = 10$.

Sums of independent Poisson random variables   Let X_1 and X_2 be two independent Poisson random variables with $X_1 \sim P(\lambda_1)$ and $X_2 \sim P(\lambda_2)$. Then their sum $Y = X_1 + X_2$ is also a Poisson random variable, with parameter $\lambda = \lambda_1 + \lambda_2$, i.e. $Y \sim P(\lambda_1 + \lambda_2)$.

It follows that if occurrences of an event over a time axis obey the Poisson assumptions listed above, the number of occurrences in any interval is Poisson distributed with a parameter proportional to the length of the interval. Thus we can write $\lambda = \alpha t$ where t is the length of the considered interval and α is the average number of occurrences per unit time.

Examples   Let X be the number of fires causing damage of more than \$100,000 in a suburb per year, with average number 4, i.e. $X \sim P(4)$:

(a) What is the probability of 6 fires in 1 year?

$$P(X = 6) = e^{-4}\frac{4^6}{6!} \tag{3.136}$$

$$= 0.1042. \tag{3.137}$$

(b) What is the probability of 12 fires in 2 years?
     Here $\lambda = 8$, since we are looking at 2 years. Thus

$$P(X = 12) = e^{-8}\frac{8^{12}}{12!} \tag{3.138}$$

$$= 0.0481. \tag{3.139}$$

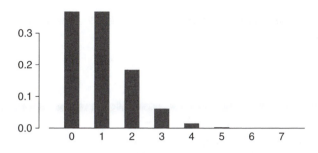

Figure 3.12 Probability function of Poisson random variable ($\lambda = 1$).

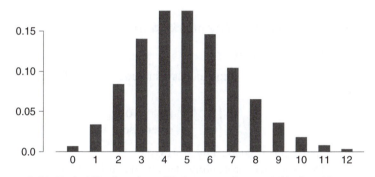

Figure 3.13 Probability function of Poisson random variable ($\lambda = 5$).

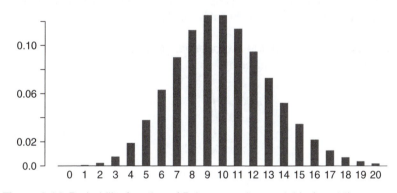

Figure 3.14 Probability function of Poisson random variable ($\lambda = 10$).

(c) What is the probability of at least 1 fire in 2 years?
    Here again $X \sim P(8)$. Thus

$$P(X \geq 1) = 1 - P(X = 0) \tag{3.140}$$

$$= 1 - e^{-8}\frac{8^0}{0!} \tag{3.141}$$

$$= 0.9997. \tag{3.142}$$

Mean and variance of a Poisson random variable The mean of a Poisson random variable with parameter λ is given by

$$E(X) = \lambda \tag{3.143}$$

and the variance by

$$\text{Var}(X) = \lambda \tag{3.144}$$

so that the standard deviation is $\sqrt{\lambda}$.

For a proof see Devore [20, pp. 135–136].

3.11.2 Continuous probability distributions

The exponential distribution

The exponential distribution is a continuous distribution with probability density function:

$$f(x) = \begin{cases} \lambda e^{-\lambda x} & \text{for } x \geq 0, \\ 0 & \text{otherwise.} \end{cases} \tag{3.145}$$

An exponential variable is denoted by $X \sim \exp(\lambda)$.

Figure 3.15 shows the probability density function of the exponential distribution.

The exponential distribution is used extensively to model the lifetime of equipment, such as batteries, light bulbs, smoke alarms or sprinklers.

Mean and variance of the exponential distribution The mean of the exponential distribution is given by

$$E(X) = \int_0^\infty x \lambda e^{-\lambda x} \, dx \tag{3.146}$$

$$= \frac{1}{\lambda}. \tag{3.147}$$

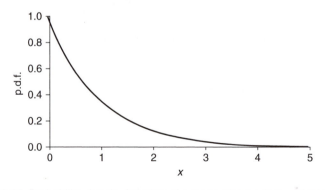

Figure 3.15 Probability density function of exponential random variable.

To calculate the variance, we first find $E(X^2)$.

$$E(X^2) = \int_0^\infty x^2 \lambda e^{-\lambda x} \, dx \qquad (3.148)$$

$$= \frac{2}{\lambda^2}. \qquad (3.149)$$

Then $\mathrm{Var}(X) = E(X^2) - [E(X)]^2 = 1/\lambda^2$, so that the standard deviation is $1/\lambda$.

Examples There are five smoke alarms in a building. Suppose that the lifetime in years of the smoke alarms, denoted by T, is exponential with parameter $\lambda = 1/5$. What is the probability that at least 2 are still functioning at the end of 8 years?

1. The probability that a given smoke alarm is still functioning after 8 years is given by

$$P(T > 8) = \frac{1}{5} \int_8^\infty e^{-t/5} \, dt \qquad (3.150)$$

$$= e^{-8/5} \qquad (3.151)$$

$$= 0.2. \qquad (3.152)$$

2. Let now X represents the number of smoke alarms functioning after 8 years. Then, using the binomial distribution.

$$P(X \geq 2) = \sum_{x=2}^{5} \binom{5}{x} 0.2^x 0.8^{5-x} \qquad (3.153)$$

$$= 1 - \sum_{x=0}^{2} \binom{5}{x} 0.2^x 0.8^{5-x} \qquad (3.154)$$

$$= 1 - 0.7373 \qquad (3.155)$$

$$= 0.2627. \qquad (3.156)$$

The normal distribution

A continuous random variable Z is said to have the *standard normal distribution* if its probability density function, denoted by $\phi(x)$, is

$$\phi(x) = \frac{1}{\sqrt{2\pi}} e^{-\frac{x^2}{2}} \quad \text{for } -\infty < x < \infty. \qquad (3.157)$$

Figure 3.16 shows the probability density function of the standard normal distribution.

The mean of a standard normal random variable is zero and the variance unity.

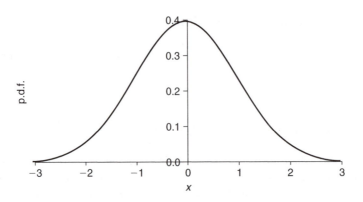

Figure 3.16 Probability density function of standard normal random variable.

The distribution function of Z is denoted by $\Phi(x)$ and is given by

$$\Phi(x) = \frac{1}{\sqrt{2\pi}} \int_{-\infty}^{x} e^{-\frac{u^2}{2}}\, du. \tag{3.158}$$

There is no explicit expression for this integral in terms of elementary functions. Its value can be obtained from tables which are given in statistical textbooks. Most computational computer programs have functions which will deliver the value of $\Phi(x)$ for all x with sufficient accuracy for all practical applications.

A random variable X is said to be *normally distributed* if it is of the form $X = \mu + \sigma Z$ where Z is a standard normal variable, $-\infty < \mu < \infty$ and $\sigma > 0$. It follows that:

1. $E(X) = \mu$.
2. $\mathrm{Var}(X) = \sigma^2$.
3. If

$$Z = \frac{X - \mu}{\sigma}, \tag{3.159}$$

then Z is standard normal.

We say that X is a normal random variable and write $X \sim N(\mu, \sigma^2)$. Clearly, since Z is a special case of X with $\mu = 0$ and $\sigma = 1$, we can write $Z \sim N(0, 1)$.

The normal distribution is the most widely used of all continuous probability distributions. It is found to be applicable to many situations. The reason is that when a random variable can be thought of as the result of the sum of a large number of independent factors, its distribution tends to be normal.

The following points should be noted:

1. The p.d.f $f(x)$ is symmetric about the mean μ.
2. $f(x)$ attains its maximum at $x = \mu$. The maximum of a p.d.f is called the *mode* of the distribution.

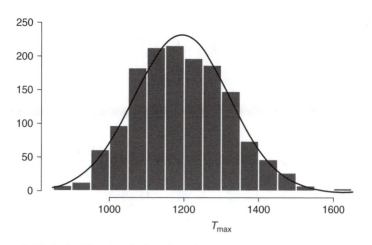

Figure 3.17 Probability density function of maximum compartment temperature.

Examples In a computer simulation of a compartment fire, with random input parameters and random forcing terms, the maximum temperature T_{max} reached (excluding smouldering fires) was found to be approximately normally distributed with mean 1194.1°C and standard deviation 127°C. There were 1484 data points. For details see Ref. [28]. Figure 3.17 shows the probability density function of the maximum temperature together with a histogram of the data.

Computing probabilities for a normal variable Let X be $\sim N(\mu, \sigma^2)$, and suppose we need to compute $P(a < X < b)$. We first notice that, by definition, $X = \mu + \sigma Z$ where Z is a standard normal variable, $\mu = E(X)$, $\sigma^2 = \text{Var}(X)$, and $\sigma > 0$. Thus, using the remark in Section 3.7.3, we see that $P(a < X < b)$ is equivalent to

$$P\left(\frac{a - \mu}{\sigma} < Z < \frac{b - \mu}{\sigma}\right) = \Phi\left(\frac{b - \mu}{\sigma}\right) - \Phi\left(\frac{a - \mu}{\sigma}\right). \qquad (3.160)$$

Note that tables of the standard normal distribution usually only give the value of $\Phi(x)$ for non-negative x. This is because, on account of the symmetry of $\phi(x)$ about the y-axis,

$$\Phi(-x) = 1 - \Phi(x). \qquad (3.161)$$

Examples In the example of Section 3.11.2 find:

1. the probability that T_{max} will be greater than 1600°C;
2. the probability that T_{max} will be less than 800°C;
3. the probability that 1000°C $< T_{max} < 1400$°C.

Solution

1.
$$P(T_{\max} > 1600) = P\left(Z > \frac{1600 - 1181.1}{176.6}\right) \quad (3.162)$$

$$= 1 - \Phi(2.372) \quad (3.163)$$

$$= 1 - 0.9912 \quad (3.164)$$

$$= 0.0088. \quad (3.165)$$

2.
$$P(T_{\max} < 800) = P\left(Z < \frac{800 - 1181.1}{176.6}\right) \quad (3.166)$$

$$= \Phi(-2.1580) \quad (3.167)$$

$$= 1 - \Phi(2.1580) \quad (3.168)$$

$$= 1 - 0.9845 \quad (3.169)$$

$$= 0.0155. \quad (3.170)$$

3.
$$P(1000 < T_{\max} < 1400) = P\left(\frac{1400 - 1181.1}{176.6} < Z \right.$$
$$\left. < \frac{1000 - 1181.1}{176.6}\right) \quad (3.171)$$

$$= \Phi(1.2395) - \Phi(-1.0255) \quad (3.172)$$

$$= \Phi(1.2395) - [1 - \Phi(1.0255)] \quad (3.173)$$

$$= 0.8924 - 0.1526 \quad (3.174)$$

$$= 0.7398. \quad (3.175)$$

Sums of independent normal variables

Theorem 3.11.1.   *Let $X_1, X_2, \ldots, X_n$ be independent normal variables with means $\mu_1, \mu_2, \ldots, \mu_n$ and standard deviations $\sigma_1, \sigma_2, \ldots, \sigma_n$. Then*

$$Y = a_0 + \sum_i a_i X_i \quad (3.176)$$

is a normal variable. In accordance with the rules for bivariate moments, we have

$$E(Y) = a_0 + \sum_i a_i \mu_i \quad (3.177)$$

$$\text{Var}(Y) = \sum_i a_i^2 \sigma_i^2. \quad (3.178)$$

For a proof see Devore [20, p. 193].

The lognormal distribution

A random variable X is said to have the *lognormal* distribution if $\ln(X)$ is normally distributed. In other words, $X = e^U$ where U is normally distributed. It is customary to use natural logarithms in this context.

The lognormal distribution has the important property that it is non-negative and is therefore very useful to model random variables which are naturally non-negative such as, for example, waiting times.

It was pointed out previously that when a random variable can be thought of as the result of the sum of a large number of independent factors, its distribution tends to be normal. It follows that when a random variable can be thought of as arising from a large number of *multiplicative* effects it will tend to have a lognormal distribution.

Although the lognormal random variable has a two-parameter distribution which is fully specified by its mean and standard deviation, it usually specified by the mean and standard deviation of its logarithm, which, as mentioned above, is normally distributed. However, it is very easy to calculate the mean and standard deviation of X in terms of those of U and vice versa. Indeed, suppose U has mean μ_U and standard deviation σ_U. Furthermore, denote the mean of X by μ_X and its coefficient of variation by CV_X. We then have the following two pairs of relations:

$$\mu_U = \ln\left(\frac{\mu_X}{\sqrt{1 + CV_X^2}}\right) \tag{3.179}$$

$$\sigma_U = \sqrt{\ln(1 + CV_X^2)}, \tag{3.180}$$

and reciprocally:

$$\mu_X = \exp\left(\mu_U + \frac{1}{2}\sigma_U^2\right) \tag{3.181}$$

$$CV_X = \sqrt{\exp(\sigma_U^2) - 1}. \tag{3.182}$$

Examples   See Hasofer and Beck [29].

A large scale Monte Carlo simulation (Chapter 5) of a one-zone fire model was carried out at the Centre for Environmental Safety and Risk Engineering (CESARE), Victoria University of Technology (VUT), Melbourne, Australia in 1997. The model used was a modified version of a model first developed by researchers at the National Research Council of Canada (NRCC) and subsequently modified by researchers at CESARE. It is now known as the NRCC/VUT fire growth model. There were 16 input parameters, which were randomly generated from given distributions.

A lognormal distribution was fitted to a sample of 1090 values of the time to untenable conditions T, in minutes. The result is shown in two plots. In the first (Figure 3.18) the time T itself is used, resulting in a lognormal distribution, while in the second (Figure 3.19) the logarithm of the time is used, resulting in a normal distribution.

Extreme value distributions

These distributions approximate the distribution of the maximum or minimum of a large random sample. The two most important ones are the *Weibull distribution* and the *Gumbel distribution*.

The Weibull distribution   The Weibull distribution approximates the distribution of the *minimum* of a large random sample from a random variable that has a lower bound. Conceptually, it is useful to think of it as the strength of a chain, the strength of whose links are the elements of the sample. Obviously, the strength of the chain is governed by the strength of the weakest link.

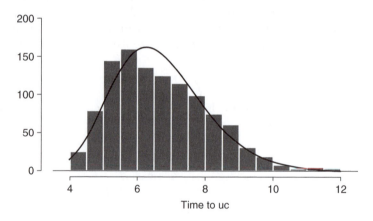

Figure 3.18 Probability density function of time to untenable conditions (uc).

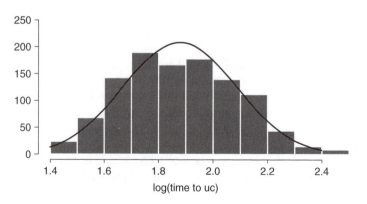

Figure 3.19 Probability density function of logarithm of time to untenable conditions (uc).

The Weibull distribution is used extensively to model material resistance, whether structural resistance, fatigue resistance or heat resistance, since resistance is inherently non-negative.

The formula for the Weibull distribution function is given by

$$F(x) = 1 - \exp\left[-\left(\frac{x - x_0}{w}\right)^{\alpha}\right] \quad \text{for } x \geq x_0 \qquad (3.183)$$

and $F(x) = 0$ for $x < x_0$, where x_0 is the lower bound of the distribution, often taken as zero.

Mean and variance   If X is a random variable having the Weibull distribution (3.183), then

$$E(X) = x_0 + w\Gamma\left(1 + \frac{1}{k}\right) \qquad (3.184)$$

and

$$\text{Var}(X) = w^2\left[\Gamma\left(1 + \frac{2}{k}\right) - \Gamma^2\left(1 + \frac{1}{k}\right)\right], \qquad (3.185)$$

where the function $\Gamma(\alpha)$ is defined by

$$\Gamma(\alpha) = \int_0^\infty e^{-x}x^{(\alpha-1)} \, dx. \qquad (3.186)$$

When α is an integer greater than 1, $\Gamma(\alpha) = (\alpha - 1)!$ and $\Gamma(1) = 1$. There is no explicit expression in terms of elementary functions when α is not an integer. Its value can be obtained from tables which are given in Mathematics handbooks. Most computational computer programs have functions which will deliver the value of $\Gamma(\alpha)$ for all α with sufficient accuracy for all practical applications.

Figure 3.20 shows a typical shape for the density function of the Weibull distribution.

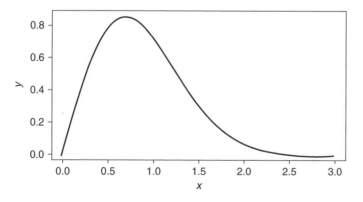

Figure 3.20 Probability density function of Weibull distribution with $x_0 = 0$ and $w = 1$.

The Gumbel distribution   The Gumbel distribution approximates the distribution of the *maximum* of a large random sample of a random variable, provided the distribution satisfies certain criteria. Often, if the criteria are not satisfied, the *logarithm* of the random variable will satisfy the criteria and its maximum will be approximated by the Gumbel distribution. For details of the criteria see Hasofer [27].

The distribution function of the Gumbel distribution is given by

$$F(x) = e^{-e^{-\frac{x-x_0}{\alpha}}} \quad \text{for } -\infty < x < \infty. \tag{3.187}$$

Mean and variance   If X is a random variable having the Gumbel distribution (3.187), then

$$E(X) = x_0 + \gamma\alpha \approx x_0 + 0.577\alpha, \tag{3.188}$$

where $\gamma \approx 0.577$ is known as "Euler's constant", and

$$\text{Var}(X) = \frac{1}{6}\pi^2\alpha^2, \tag{3.189}$$

where π is as usual 3.1416.

Figure 3.21 shows the shape of the density function of the Gumbel distribution with $x_0 = 0$ and $\alpha = 1$.

Example   The maximum yearly fire loss for industrial fires in the UK for the years 1966 to 1972, in units of one thousand pounds was recorded for seven different occupations. The buildings were either single storey or multistorey and unsprinklered. The occupations covered were as follows:

1. Textiles
2. Timber and furniture
3. Paper printing and publishing
4. Chemical and allied industries
5. Manufacture engineering and electrical goods industries

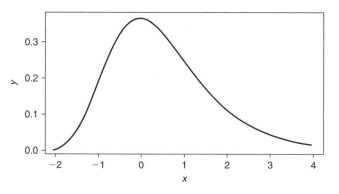

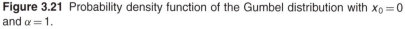

Figure 3.21 Probability density function of the Gumbel distribution with $x_0 = 0$ and $\alpha = 1$.

6. Wholesale distributive trades
7. Retail distributive trades.

Thus the size of the sample was $14 \times 7 = 98$.

Figure 3.22 presents a histogram of the (natural) logarithms of the data together with the density function of a Gumbel distribution with parameters $x_0 = 5.34$ and $\alpha = 1.35$.

3.12 Estimation

It is often necessary, when a random sample from a distribution is given, to *estimate* the parameters of the parent distribution from the given random sample. The most frequently used method is called the *estimating equation* method. Let the parameters of the distribution be $\alpha_1, \alpha_2, \ldots, \alpha_n$. The method involves finding n functions of the sample values and the parameters:

$$g_i(\alpha_1, \alpha_2, \ldots, \alpha_n; X_1, X_2, \ldots, X_n), \quad i = 1, \ldots, n, \qquad (3.190)$$

such that $E(g_i) = 0$ for all i. We then solve the n equations $g_i = 0$ for the α_i in terms of the X_i. The estimated values are denoted by $\hat{\alpha}_1, \hat{\alpha}_2, \ldots, \hat{\alpha}_n$.

A frequently used set of estimating equations is the so-called *method of moments*, in which we equate the moments of the distribution to the corresponding sample moments, expressed in a form that ensures that their expectation is equal to the distribution moment.

The sample mean, denoted by $\bar{X}$, is defined by

$$\bar{X} = \frac{1}{n}(X_1 + X_2 + \cdots + X_n). \qquad (3.191)$$

The sample variance, denoted by s^2, is defined by

$$s^2 = \frac{1}{(n-1)}\left[(X_1 - \bar{X})^2 + (X_2 - \bar{X})^2 + \cdots + (X_n - \bar{X})^2\right]. \qquad (3.192)$$

The definitions are chosen so that $E(\bar{X}) = E(X)$ and $E(s^2) = \mathrm{Var}(X)$.

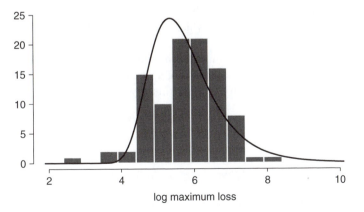

Figure 3.22 Probability density function of logarithm of maximum yearly fire loss.

When there is only one parameter, we can use as estimating equation $E(X) = \bar{X}$, since clearly $E[E(X) - \bar{X}] = 0$. When there are two parameters, we can similarly use as a second estimating equation $\text{Var}(X) = s^2$.

Examples

1. Let $\bar{X}$ be the sample mean of a random sample from the exponential distribution (3.146). Then, since $E(X) = 1/\lambda$, the estimated value of the parameter λ, called the *estimator* of λ, is $\hat{\lambda} = 1/\bar{X}$.

2. Let $\bar{X}$ be the sample mean and s^2 the sample variance of a random sample from the binomial distribution (3.129). Here we need two estimating equations, since we have the two parameters v and p. We choose as estimating equations

$$vp = \bar{X} \tag{3.193}$$

$$vp(1 - p) = s^2 \tag{3.194}$$

solving for v and p we obtain

$$\hat{p} = 1 - \frac{s^2}{\bar{X}} \tag{3.195}$$

$$\hat{v} = \frac{\bar{X}^2}{\bar{X} - s^2}. \tag{3.196}$$

3. In the case of the normal distribution, we simply have

$$\hat{\mu} = \bar{X} \tag{3.197}$$

$$\hat{\sigma}^2 = s^2. \tag{3.198}$$

3.13 Confidence interval

It is often useful, when estimating a parameter from a distribution, to have some measure of the precision of the estimate. This precision clearly depends on the variability of the estimate, which can be measured through the standard deviation.

It is often the case that the estimator of a parameter is approximately normally distributed about the true value of the parameter. Let the estimate be $\hat{\alpha}$, with mean α and standard deviation σ. If $\hat{\alpha}$ is approximately normally distributed, we have

$$P(\alpha - 1.96\sigma \le \hat{\alpha} \le \alpha + 1.96\sigma) = 0.95. \tag{3.199}$$

This can be rewritten

$$P(\hat{\alpha} - 1.96\sigma \le \alpha \le \hat{\alpha} + 1.96\sigma) = 0.95. \tag{3.200}$$

The last equation can be interpreted to mean that if the estimation is repeated a large number of times, the true value α will fall inside the interval

$(\hat{\alpha} - 1.96\sigma, \hat{\alpha} + 1.96\sigma)$ 95% of the time. The interval $(\hat{\alpha} - 1.96\sigma, \hat{\alpha} + 1.96\sigma)$ is called a 95% *confidence interval*.

If the true value of σ is unknown, it is replaced by its estimator $\hat{\sigma}$.

Examples   Let $X_1, \ldots, X_n$ be a random sample of size n of Bernoulli variables (Section 3.11.1) with probability of success p. We wish to find a 95% confidence interval for p. Let the observed number of successes be n_1. Since $E(n_1) = np$, we take $\hat{p} = n_1/n$ as the estimator of p. The standard deviation of n_1 is $\sqrt{np(1-p)}$. Therefore, the estimated standard deviation of $\hat{p}$ is

$$\hat{\sigma} = \sqrt{\frac{\hat{p}(1-\hat{p})}{n}}. \tag{3.201}$$

It can be shown that, for large n, $\hat{p}$ is approximately normally distributed. Thus the 95% confidence interval for p can be written as

$$\hat{p} \pm 1.96\sqrt{\frac{\hat{p}(1-\hat{p})}{n}}. \tag{3.202}$$

3.14 Regression

3.14.1 An example

Table 3.6 gives the yearly direct fire loss for Japan as a percentage of gross domestic product for the period 1984–1993 (from Ramachandran [54]).

In Figure 3.23 the percentage fire loss is plotted against the year. It is clear that there is a general trend towards a reduction of the percentage and that the

Table 3.6 Direct fire loss for Japan as a percentage of gross domestic products

Year	1984	1985	1986	1989	1990	1991	1992	1993
Percentage fire loss	0.18	0.20	0.16	0.11	0.12	0.07	0.10	0.08

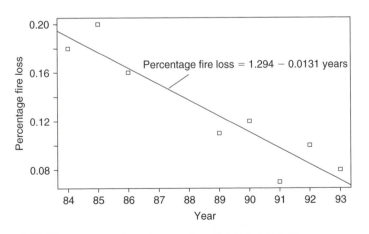

Figure 3.23 Linear regression of percentage fire loss on year.

trend looks quite linear. If we let y be the percentage fire loss and x be the last two digits of the year, i.e. counting time from the year 1900, we can expect to represent the trend by an equation of the form.

$$y = a + bx. \qquad (3.203)$$

Such a representation is called the *linear regression* of y on x.

3.14.2 Evaluation of coefficients

Suppose we are given values $(x_1, \ldots, x_n)$ of the variable x and corresponding values $(y_1, \ldots, y_n)$ of the variable y. The most widespread method for fitting a regression line $y = a + bx$ to the data is called *least-square fitting*. It consists in finding values of a and b that will minimize the *objective function*:

$$Q = \sum_{i=1}^{n} (y_i - a - bx_i)^2. \qquad (3.204)$$

Equating the partial derivatives of Q with respect to a and b to zero we obtain the two equations:

$$na + b \sum_{i=1}^{n} x_i = \sum_{i=1}^{n} y_i \qquad (3.205)$$

$$a \sum_{i=1}^{n} x_i + b \sum_{i=1}^{n} x_i^2 = \sum_{i=1}^{n} x_i y_i. \qquad (3.206)$$

It is easy to show that the solution of these equations is

$$b = \frac{\sum_{i=1}^{n} (x_i - \bar{x})(y_i - \bar{y})}{\sum_{i=1}^{n} (x_i - \bar{x})^2} \qquad (3.207)$$

and

$$a = \bar{y} - b\bar{x}, \qquad (3.208)$$

where $\bar{x}$ is the mean of the x_i,

$$\bar{x} = \frac{1}{n} \sum_{i=1}^{n} x_i \qquad (3.209)$$

and $\bar{y}$ is the mean of the y_i.

For the data of Table 3.6, the values of a and b are found to be 1.294 and 0.0131, respectively.

3.14.3 Analysis of variance

The analysis of variance technique allows us to determine how much of the variability of y is explained by the regression line.

Let $\hat{y} = a + bx$, where a and b are the values obtained by least-square fitting. $\hat{y}$ is called the *predicted value* of y for the given value of x. We then have

$$\sum_{i=1}^{n} (y_i - \bar{y})^2 = \sum_{i=1}^{n} (\hat{y}_i - \bar{y})^2 + \sum_{i=1}^{n} (y_i - \hat{y}_i)^2. \qquad (3.210)$$

This achieves *partitioning* of the total "corrected sum of squares of y" into two components. We write this symbolically as

$$\text{SST} = \text{SSR} + \text{SSE} \qquad (3.211)$$

The first component on the right is called the *regression sum of squares* and it reflects the amount of variation in the y-values explained by the model. The second component reflects variation about the regression line.

Let R be the correlation coefficient (3.10.1) between the y_i and the corresponding values $\hat{y}_i$ predicted by the model. It is given by the formula:

$$R = \frac{\sum_{i=1}^{n} (y_i - \mu)(\hat{y}_i - \hat{\mu})}{\{[\sum_{i=1}^{n} (y_i - \mu)^2][\sum_{i=1}^{n} (\hat{y}_i - \hat{\mu})^2]\}^{1/2}}, \qquad (3.212)$$

where $\mu = (1/n) \sum_{i=1}^{n} y_i$ and $\hat{\mu} = (1/n) \sum_{i=1}^{n} \hat{y}_i$.

It can be shown that

$$R^2 = \frac{\text{SSR}}{\text{SST}}. \qquad (3.213)$$

In other words, the square of the correlation between the true values of y and the predicted values of $\hat{y}$ is equal to the proportion of the variation in y explained by the regression. Thus R^2 can be taken as a measure of the goodness of fit of the data to the model.

For the fire loss data given above, the value of R^2 is 0.87, so that 87% of the variation is explained by the regression.

3.14.4 Multiple linear regression

Suppose now that the dependent variable y is a function of k independent variables $x_1, \ldots, x_k$. The equation $y = f(x_1, \ldots, x_k)$ defines a surface in the k-dimensional space of $x_1, \ldots, x_k$. It is called a *response surface*. Let $\mathbf{x} = (x_1, \ldots, x_k)$.

We try to fit the model

$$y = b_0 + b_1 x_1 + \cdots + b_k x_k. \qquad (3.214)$$

Let $\mathbf{b} = (b_1, \ldots, b_k)^{\text{T}}$.

The given data are $\mathbf{y} = (y_1, \ldots, y_n)^{\text{T}}$ and

$$\mathbf{X} = \begin{pmatrix} 1 & x_{11} & x_{12} & \cdots & x_{1k} \\ 1 & x_{21} & x_{22} & \cdots & x_{2k} \\ \vdots & \vdots & \vdots & \vdots & \vdots \\ 1 & x_{n1} & x_{n2} & \cdots & x_{nk} \end{pmatrix}, \qquad (3.215)$$

where each row represents a data point and the y_i are the corresponding values of y. Least-square fitting now attempts to minimize the objective function:

$$Q = (\mathbf{y} - \mathbf{Xb})^{\mathrm{T}}(\mathbf{y} - \mathbf{Xb}). \qquad (3.216)$$

It turns out that $\mathbf{b}$ is the solution of the vector equation:

$$(\mathbf{X}^{\mathrm{T}}\mathbf{X})\mathbf{b} = \mathbf{X}^{\mathrm{T}}\mathbf{y}. \qquad (3.217)$$

Computer programs for performing multiple regression analysis are available in most scientific or engineering computing packages, e.g. Mathcad User's Guide [45].

3.14.5 Analysis of variance for multiple linear regression

Similarly to simple linear regression let the predicted value of y be $\hat{y} = b_0 + b_1 x_1 + \cdots + b_k x_k$.

The analysis of variance formula

$$\sum_{i=1}^{n} (y_i - \bar{y})^2 = \sum_{i=1}^{n} (\hat{y}_i - \bar{y})^2 + \sum_{i=1}^{n} (y_i - \hat{y}_i)^2 \qquad (3.218)$$

continues to hold. Retaining the previous notation, we have

$$\mathrm{SST} = \mathrm{SSR} + \mathrm{SSE} \qquad (3.219)$$

and letting R denote the correlation coefficient of y_i and $\hat{y}_i$, given by the formula (3.212), we still have, as for the simple regression

$$R^2 = \frac{\mathrm{SSR}}{\mathrm{SST}}. \qquad (3.220)$$

i.e. R^2 represents the fraction of the total variability of y that is explained by the regression.

3.14.6 Example 1

In Magnusson *et al.* [43] an empirical equation is derived for the time S in seconds taken by the smoke layer in a compartment to come down to 2 m below the ceiling for the computer fire model CFAST [11].

The proposed empirical equation is

$$S = C\alpha^p H^q A^r, \qquad (3.221)$$

where α is the fire growth rate in kW/s^2, H is the ceiling height in m and A is the floor area in m^2. The parameters C, p, q and r are to be determined by fitting the model to CFAST data.

There were 39 runs of CFAST available with the values of the input variables chosen from Table 3.7.

At first glance, it does not appear that the fitting of equation (3.221) is a linear regression problem. However, if we take logarithms, equation (3.221) becomes

$$\log(S) = \log(C) + p\log(\alpha) + q\log(H) + r\log(A) \tag{3.222}$$

and the parameters C, p, q, r can be obtained from a linear regression. The values found were $C = 1.67$, $p = -0.36$, $q = 0.44$ and $r = 0.54$. Figure 3.24 shows a plot of the smoke filling time predicted by regression against the smoke filling time calculated from CFAST.

Actually, Magnusson *et al.* minimize a slightly different function, which they call the "squared relative error", namely

$$Q = \sum_i \left(\frac{C\alpha_i^p H_i^q A_i^r - S_i}{C\alpha_i^p H_i^q A_i^r} \right)^2. \tag{3.223}$$

It can be shown that when the relative error is small, as is the case here, the results of the two approaches are nearly the same.

Table 3.7 Input values for the calculation of smoke filling time

Parameters				
Floor area (m²)	200	500	800	1600
Ceiling height (m)	3	5	8	
Fire growth rate (kW/s²)	0.001	0.005	0.01	0.02

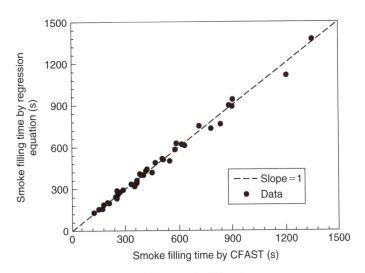

Figure 3.24 Regression analysis for smoke filling time.

3.14.7 Polynomial regression

It may happen that the experimenter feels that a more appropriate model than linear regression would be a *polynomial regression*. For example, suppose that y depends on just two variables x_1 and x_2. However, the experimenter feels that the model

$$y = b_0 + b_1 x_1 + b_2 x_2 \qquad (3.224)$$

is too crude, and would like to add quadratic components to both variables. In other words, the experimenter would like to fit the model:

$$y = b_0 + b_1 x_1 + c_1 x_1^2 + b_2 x_2 + c_2 x_2^2. \qquad (3.225)$$

Evaluating the coefficients poses no new difficulties. All that is required is to consider x_1, x_1^2, x_2, x_2^2 as the new independent variables, numbering now 4.

For examples of polynomial regression see Chapter 11.

4

Beta reliability index

4.1 The multivariate normal distribution

Let $U = (U_1, U_2, \ldots, U_n)$ be a vector of independent standard normal variables of length n and let A be an $n \times n$ non-singular matrix. Furthermore, let $\mu = (\mu_1, \mu_2, \ldots, \mu_n)$ be a vector of constants of length n. Let

$$X = AU + \mu. \tag{4.1}$$

Then X is said to be a vector of *multinormal variables*. Following the method of Section 3.10.2, we find

$$E(X) = \mu \tag{4.2}$$

$$\mathrm{Var}(X) = AE(UU^T)A^T \tag{4.3}$$

$$= AA^T \tag{4.4}$$

since $E(UU^T) = I$, where I is the $n \times n$ unit matrix.

Conversely, let X be a multivariate normal vector with mean μ.

Let $\Sigma = E[(X - \mu)(X - \mu)^T]$ be the (non-singular) variance–covariance matrix of X.

Suppose that we can find a decomposition of Σ in the form

$$\Sigma = AA^T \tag{4.5}$$

where A is a non-singular square matrix. Then $U = A^{-1}(X - \mu)$ will be a vector of independent standard normal variables. For a proof see Ref. [40, p. 347]. This is called a process of *standardization* of X.

One well-known form of the decomposition (4.5) is the *Choleski* decomposition, where A is chosen to be a lower triangular matrix. For details see Ref. [63]. Most computer programs have an implementation of the Choleski algorithm.

It is also clear that any rotation of the vector U will not affect the representation (4.1). For let $U = HV$, where H is an orthogonal matrix, i.e. $HH^T = I$. Then $V = H^T U$, $E(V) = 0$ and

$$\mathrm{Var}(V) = H^T E(UU^T)H = I. \tag{4.6}$$

Moreover $X = A^*V + \mu$, where $A^* = AH$. Thus it is possible to choose the direction of the axes in U space in such a way that the axis U_1 points in any desired direction.

4.2 The limit state formulation of safety

We assume that safety in the considered situation depends on k variables $X_1, X_2, \ldots, X_k$ which are called the *basic variables* of the problem. We denote them by the vector $X = (X_1, X_2, \ldots, X_k)$.

Safety is defined by a *limit state surface* in the space of the k basic variables, represented mathematically by the limit state equation $G(\mathbf{X}) = \mathbf{0}$. Safety is achieved if $G(\mathbf{X}) > \mathbf{0}$, while if $G(\mathbf{X}) \leq \mathbf{0}$ the situation is unsafe.

It should be noted that the choice of basic variables is not unique. Theoretically, there is an infinity of possible choices. However, in engineering problems, there is usually only one natural way to chose the basic variables, apart from the choice of units, which is of course arbitrary.

Example   Suppose that toxic fumes from a fire contain X_1 and X_2 parts per thousand of two toxic chemicals. So here $\mathbf{X} = (X_1, X_2)$. It has been experimentally determined that a 5-min exposure to the fumes will be unsafe if

$$G(\mathbf{X}) = 140 - 10X_1 - X_2 - 0.1X_2^2 \leq \mathbf{0}. \tag{4.7}$$

Thus, if $X_1 = 2$ and $X_2 = 10$ the 5-min exposure to the fumes will be safe, while if $X_1 = 10$ and $X_2 = 20$ the exposure is unsafe.

4.3 The case of one variable

Suppose that there is only one basic variable X and that the limit state equation is $G(X) = X - x_0$, so that safety is achieved if $X < x_0$.

Suppose now that X is a random variable that is approximately normally distributed with mean μ and standard deviation σ. We can then write $X = \sigma U + \mu$, where U, the corresponding standardized random variable, has zero mean and unit standard deviation.

Replacing X in terms of U in $G(X)$, we obtain a new limit state equation:

$$G^*(U) = G(\sigma U + \mu) = \sigma U + \mu - x_0 \tag{4.8}$$

and safety is achieved if

$$U < \frac{x_0 - \mu}{\sigma} = \beta \text{ say.} \tag{4.9}$$

We now use the fact that X and therefore U are approximately normally distributed. Thus the probability of safety is given by $P(U < \beta) \approx \Phi(\beta)$, where $\Phi(x)$ is the distribution function of the standard normal variable. This probability is usually very nearly unity, so it is usual to quote the *probability of failure* (i.e. the probability of failing to achieve safety), which is $P(X \geq 0) = 1 - \Phi(\beta)$.

In the beta reliability index method we take β to be a measure of safety. The larger β, the safer the situation. If X is exactly normally distributed, we have the following relationship between β and $P(X < x_0)$ (Table 4.1).

Table 4.1 Probability of failure

β	2	3	4
$P(X \geq x_0)$	0.023	0.0013	0.000032

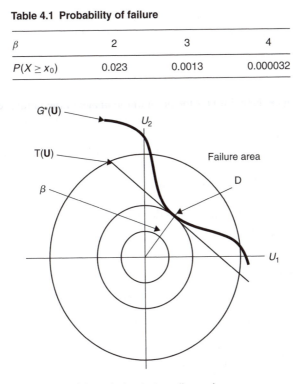

Figure 4.1 Illustration of the β index in two dimensions.

4.4 The multivariate case

The univariate definition of the beta reliability index can be extended to the multivariate case as follows.

Let **X** be the vector of basic variables and $G(\mathbf{X})$, the limit state equation. Using the results of Section 4.1, we write $\mathbf{X} = \mathbf{A}\mathbf{U} + \boldsymbol{\mu}$ where **U** is a set of uncorrelated standard normal variables. The image of the limit surface $G(\mathbf{X}) = 0$ in the **U** space is $G^*(\mathbf{U}) = \mathbf{G}(\mathbf{A}\mathbf{U} + \boldsymbol{\mu}) = \mathbf{0}$.

The reliability coefficient β is now defined as the distance from the origin to the limit surface $G^*(\mathbf{U}) = \mathbf{0}$ in the **U** space. This is illustrated for the two-dimensional case in Figure 4.1.

It is to be noted that the sphere C centred at the origin with radius β just touches the image of the limit state surface $G^*(\mathbf{U}) = \mathbf{0}$ at some point D, which is known as the "design point". At that point the tangent hyperplane T to the sphere C also touches the limit surface $G^*(\mathbf{U}) = \mathbf{0}$ and is perpendicular to the line from the origin to D.

The rationale for the choice of β is as follows: suppose the elements of **X** are jointly normally distributed. Then the elements of **U** are independent standard normal variables. As was pointed out in Section 4.1, it is always possible to choose the directions of the axes in **U** space so that U_1 points in any desired direction (if necessary an appropriate rotation should be performed). We choose the direction of U_1 to be along the line from the origin to D. It then turns out that the probability of the region in **U** space beyond the tangent hyperplane T (which is perpendicular to U_1) is given by

$$P(\mathbf{U} > \beta) = 1 - \Phi(\beta) = \Phi(-\beta). \tag{4.10}$$

Moreover, since the point D is the point of the image of the failure region in **U** space nearest to the origin, most of the probability of failure is concentrated near it. But it is precisely in the neighbourhood of the point D that the tangent plane T approximates best the surface $G^*(\mathbf{U}) = 0$. So unless the limit state surface is extremely irregular near the design point, $\Phi(-\beta)$ will provide a very good approximation to the failure probability.

The design point can be found by using any of the constrained optimization algorithms available. One algorithm that works in most situations is the use of Lagrange multipliers.

4.5 Example

Consider the limit state function given in Section 4.2, namely

$$G(X_1, X_2) = 140 - 10X_1 - X_2 - 0.1X_2^2. \tag{4.11}$$

Let X_1 and X_2 be jointly normal variables. Let the mean of X_1 be $\mu_1 = 2$ and the mean of X_2 be $\mu_2 = 10$. Furthermore let the variance–covariance matrix of X_1 and X_2 be

$$\Sigma = \begin{pmatrix} 1.06 & 5 \\ 5 & 100 \end{pmatrix}. \tag{4.12}$$

The matrix **A** corresponding to Σ is given by

$$\mathbf{A} = \begin{pmatrix} 0.9 & 0.5 \\ 0 & 10 \end{pmatrix}. \tag{4.13}$$

In other words

$$\begin{pmatrix} X_1 \\ X_2 \end{pmatrix} = \begin{pmatrix} 2 \\ 10 \end{pmatrix} + \begin{pmatrix} 0.9 & 0.5 \\ 0 & 10 \end{pmatrix} \begin{pmatrix} U_1 \\ U_2 \end{pmatrix}. \tag{4.14}$$

From equation (4.1) we obtain

$$X_1 = 2 + 0.9U_1 + 0.5U_2 \tag{4.15}$$

$$X_2 = 10 + 10U_2. \tag{4.16}$$

It follows that

$$G^*(\mathbf{U}) = 100 - 9U_1 - 40U_2 - 10U_2^2. \tag{4.17}$$

The design point, i.e. the point on the curve $G(\mathbf{U}) = 0$ nearest to the origin is (0.2085, 1.7165) and the distance is $\beta = 1.729$. This corresponds to a probability of failure of $\Phi(-1.729) = 0.042$.

5

The Monte Carlo method

5.1 Introduction

There are many situations in probabilistic risk analysis when there is no analytic algorithm that will evaluate the required probabilities. Alternatively, the available algorithm is extremely complex and can only be carried out at great expense of effort and computer time.

An alternative method is known as Monte Carlo simulation. It depends on the fact that the histogram of a large random sample approximates the probability function of the underlying random variable.

Suppose that the output variable required to carry out the risk analysis, denoted by Y, is given as a function of a vector $\mathbf{X}$ of underlying variables: $\mathbf{X} = (X_1, \ldots, X_n)$ in the form

$$Y = f(\mathbf{X}). \tag{5.1}$$

In the Monte Carlo method, a random sample of size N of the vector of underlying variables $\mathbf{X}_1, \ldots, \mathbf{X}_N$ is generated. Each such vector is called a *realization* of the vector $\mathbf{X}$. To each realization there corresponds a value of the output variable Y. Thus we obtain a sample of size N from the output variable Y. Provided N is chosen appropriately large, the histogram of Y will approximate its distribution as closely as required.

Example As an example, we shall consider a floor of a building consisting of four compartments numbered 1 to 4 and we want to study fire spread from compartment 1 to compartment 4.

For the purpose of investigating fire spread from one compartment to an adjacent one, a building can be represented by a network. Each compartment is represented by a vertex, and vertices are connected by an edge if there is between the two corresponding compartments a direct path that the fire can use. To each edge we assign a random variable representing the time taken by the fire to spread from the compartment at the start of the edge to the compartment at the other end. Let us denote the random variable assigned to edge number r, which links vertex i to vertex j, by T_r.

The floor configuration as well as the corresponding network are shown in Figure 5.1.

Suppose that we are given the probability distributions of the various times taken by the fire to spread from one room to the next. As shown in the figure, the time for the fire to spread from room 1 to room 2 (or vice versa from room 2 to room 1) is denoted by T_1 and so on. We shall assume, for simplicity, that

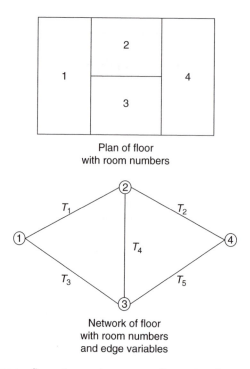

Plan of floor
with room numbers

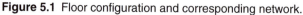

Network of floor
with room numbers
and edge variables

Figure 5.1 Floor configuration and corresponding network.

these times are independent of each other. We are interested in determining the probability distribution of the time taken by the fire to spread from room 1 to room 4.

There are four paths that the fire can take: $(1, 2, 4), (1, 3, 4), (1, 2, 3, 4)$, $(1, 3, 2, 4)$. The corresponding times, which we shall denote by U_1, U_2, U_3, U_4, are given by

$$
\begin{aligned}
U_1 &= T_1 + T_2 \\
U_2 &= T_3 + T_5 \\
U_3 &= T_1 + T_4 + T_5 \\
U_4 &= T_3 + T_4 + T_2.
\end{aligned}
\tag{5.2}
$$

The time W taken by the fire to spread from room 1 to room 4 will clearly be the minimum of these 4 times:

$$
W = \min(U_1, U_2, U_3, U_4).
\tag{5.3}
$$

There is no easy way to obtain analytically the distribution of W. On the other hand, a Monte Carlo simulation is extremely easy to carry out and will provide as good an approximation as may be required, if a sufficiently large sample is used.

We shall assume that all five random variables T_1, T_2, T_3, T_4, T_5 have a lognormal distribution with the same parameters. We recall that a lognormal

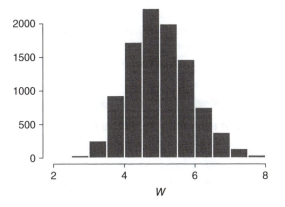

Figure 5.2 Histogram of *W*.

random variable is usually specified by the mean and standard deviation of its (natural) logarithm, which has a normal distribution. Here we shall assume that the mean and standard deviation of the logarithms are 1 and 0.3, respectively.

The Monte Carlo simulation consists in constructing independent random samples of equal size from T_1, T_2, T_3, T_4, T_5 and then constructing from them the U_i and finally *W*, using formulae (5.2) and (5.3). The following results are based on samples of size 10,000:

1. The histogram of the sample from *W* is given by Figure 5.2.
2. The mean of the (natural) logarithm of *W* is 1.60.
3. The standard deviation of the logarithm of *W* is 0.18.

Of course, what is mostly of interest in fire engineering is the probability that the fire will spread quickly, which might cause many casualties. It is very easy to obtain such probabilities. For example, if we want to know the probability that the fire will spread from room 1 to room 4 in less than 3.5 min, all we need to do is count the number of values of *W* that are less than 3.5. In the simulation just described, the number was 298, which corresponds to a probability of $298/10,000 \approx 0.03$.

5.2 The confidence interval for a Monte Carlo simulation

Suppose that a Monte Carlo simulation of size *N* is carried out to determine the probability of some subset *A* of the output space. Suppose that the output of *n* simulations is in *A*. The Monte Carlo estimator of p_A, the probability of *A*, is $\hat{p}_A = n/N$. Now each realization of the input can be thought of as Bernoulli trial (Section 3.11.1) with probability of success p_A. Following the procedure of the example of Section 3.13, we can calculate a 95% confidence interval for the estimator $\hat{p}_A$ as

$$\hat{p}_A \pm 1.96\sqrt{\frac{\hat{p}_A(1-\hat{p}_A)}{N}}. \tag{5.4}$$

Clearly, the larger N, the shorter the confidence interval and the greater the precision of the estimator. In fact, the length of the confidence interval varies inversely as the square root of the number of simulations.

For the example of the preceding section, the 95% confidence interval turns out to be $(0.0265, 0.0331)$.

5.3 Confirmation of reliability by Monte Carlo

Monte Carlo simulation can be used to confirm the probability of failure obtained from the reliability index. The easiest method is to work in the $\mathbf{U}$ space. A set of N realizations of the standardized vector $\mathbf{U}$ is generated and each $\mathbf{U}$ is tested to determine whether it falls in the safe region or the failure region. The probability of the failure region is then estimated by n/N, where n is the number of vectors $\mathbf{U}$ that fall in the failure region.

Example   Consider the example in Section 4.5. The limit state curve is given by

$$G^*(\mathbf{U}) = 100 - 9U_1 - 40U_2 - 10U_2^2. \tag{5.5}$$

A set of 100,000 realizations of the vector $\mathbf{U}$ was generated and the value of G^* calculated for each realization. There were 4155 realizations for which $G^*(\mathbf{U}) < 0$. So the Monte Carlo estimator of the probability of failure is 0.0416 and the 95% confidence interval is $(0.0403, 0.0428)$. The probability of failure obtained from the β reliability index is

$$\Phi(-1.729) = 0.042, \tag{5.6}$$

which falls inside the Monte Carlo confidence interval.

6

Event and fault trees

6.1 Introduction

Risk analysis is normally based on identification and probabilistic evaluation of failure scenarios. Ideally, one would like to identify all of them in a complete and systematic way. A failure scenario can be described by a sequence of events, some of them external (e.g. fire spreading from a neighbouring building) and some of them internal to the system (e.g. failure of a valve to open). These events can be displayed in different graphical forms. One convenient structure to encompass a family of scenarios is the structure of a tree representing a sequential progression of branching points at which several possibilities can be envisioned. Two particular structures of tree are of interest in risk analysis: event trees and fault trees.

6.2 Steps in risk analysis

6.2.1 Loss scenario development

A loss scenario represents the sequence of events that can result in a fire. The scenario development process must be:

1. Sequentially structured in a time-related manner
2. Credible in terms of realistic incident outcomes
3. Contain sufficient information to allow the risk analysis team to quantify the scenario.

The identification of the scenario sequences should be done by using event trees.

6.2.2 Exposure assessment

The general modelling approach for exposure assessment involves the following steps:

1. Determine the vulnerable targets. Vulnerability is defined in terms of the potential failure limits of the target subsystems or components when exposed to the considered fire scenario. The main targets in building fires are the occupants, the structure and the property stored in the building.
2. Model the spread of fire for the particular scenario under consideration.

3. Model the response of the fire barriers, the detection and fire suppression systems, and the occupant movements.
4. Evaluate the risk to life and to property.

6.2.3 Fire protection systems

Fire protection systems of primary interest for risk analysis include:

1. Detection systems, e.g. smoke detectors, radiant and thermal energy detectors.
2. Emergency control systems, e.g. emergency shutdown of electricity and gas, emergency startup of safety systems such as smoke removal systems.
3. Automatic suppression systems, i.e. sprinklers.
4. Fire barriers.
5. Manual intervention, e.g. manual activation of fire alarm, use of fire extinguishers and fire brigade intervention.

Risk evaluation of the fire protection systems is carried out by using fault trees.

6.3 Event trees

An event tree, also called probability tree, is a useful graphical way of representing the dependence of events. It is formed of a sequence of random variables (usually discrete) or events sets that can be associated with random variables. The branching point at which a new variable is introduced in the tree is called a node. Each node is followed by the possible realizations of this new variable, and their probabilities conditional on values of previous random variables in the tree. The most common way of constructing an event tree is to use deductive logic ("forward logic"); i.e., starting with an initiating event, lay out all the possible sequences of following events, and determine the outcome of each considered sequence. Because the probability of each event is displayed conditional on the occurrence of events that precede it in the tree, the joint probability of the simultaneous occurrence of events that constitute a sequence (i.e. a "scenario") is found by multiplication.

The simplest way to understand event trees is by means of an example. Consider again the example of Section 3.4. We can represent the dependence relations between the four events DC, WC, DO, WO by an event tree, as shown in Figure 6.1.

The figure above each event represents the probability of the event conditional on the previous events in the tree. For example, the figure 0.25 above the event WC at the right represents the probability of WC conditional on DC. The figure 0.1 below WC at the right represents the probability of the branch $DC - WC$, i.e. it represents the event $DC \cap WC$. And since, by definition, $P(DC \cap WC) = P(DC)P(WC \mid DC)$, we see that the probability of a branch is obtained by multiplying the probabilities along the branch. In the case of the

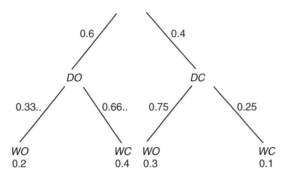

Figure 6.1 Event tree for closing and opening of doors and windows.

right branch we have been looking at, the final figure, 0.1, is obtained by multiplying 0.4 by 0.25. Finally, the total probability of *WC* is obtained by adding the probabilities of all the branches ending in *WC*. For the given figure, we have $P(WC) = 0.4 + 0.1 = 0.5$. The event tree constitutes a graphical representation of the theorem of total probability.

6.4 Fault trees

6.4.1 Aim

The goal of fault tree construction is to model the system conditions that can result in the occurrence of some designated undesired event, called the top event. The fault tree graphically and logically represents the various combinations of possible events, both fault and normal, occurring in a system that lead to the undesired event.

6.4.2 Fault events

While an event tree is formed of a sequence of random variables, a fault tree is formed of events described by binary variables, i.e. variables that take the value 1 if the event occurs and 0 if it does not. These events are related by logical functions, essentially OR and AND. Graphically, these logical functions are represented by *Boolean gates*. The OR gate describes a situation where the output event of the gate will occur if one or more of the input events occur. The AND gate describes the logical operation that requires the simultaneous occurrence of all input events to produce the output event. The fault tree is so structured that the sequences of events that lead to the top event are shown below the top event and are logically related to the undesired event by OR and AND gates. The input events to each logic gate may also be outputs of other logic gates at a lower level. The standard symbols for events and gates are shown in Figure 6.2. The events that are not outputs of other events are called *basic events*.

As an example of AND and OR gates, consider the simple series circuit controlling a motor, as shown in Figure 6.3.

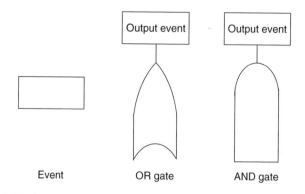

Event   OR gate   AND gate

Figure 6.2 Fault tree analysis symbols.

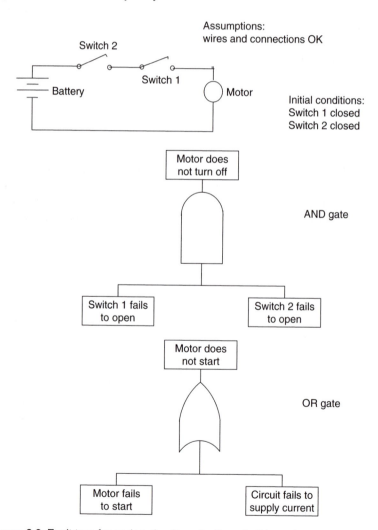

Figure 6.3 Fault tree for series circuit controlling electric motor.

The first fault tree in Figure 6.3, illustrating an AND gate, examines one type of top event, namely "motor not turning off". The second fault tree, illustrating an OR gate, examines another type of top event, namely "motor does not start". It is assumed that the wires and connectors do not contribute to system failure.

6.4.3 Evaluation of the fault tree

Evaluation of the fault tree can be either qualitative or quantitative (determination of the probability of the top event). If the system is found to be inadequate, it must be upgraded by first identifying critical events (such as component failures) that significantly contribute to the top event. Changes are then made, subject to cost constraints, contractual requirements and other factors to reduce the effect of the critical events. Finally the new fault tree is evaluated.

6.4.4 Qualitative evaluation of the fault tree

Qualitative evaluation of the fault tree is carried out by determining the *minimum cut sets*. A cut set is a set of basic events whose simultaneous occurrence will cause the top event to occur. A cut set is said to be minimal if it cannot be reduced and still ensure the occurrence of the top event. A listing of minimal cut sets is useful for design purposes, to determine the "weakest links" of the system.

For small fault trees minimal cut sets can be determined by inspection. However, with large, complex systems, which can involve hundreds of fault events, recourse must be had to systematic algorithms to determine the minimal cut sets. The reader is referred to the specialized literature, e.g. Ref. [36] on the subject for a full description of the algorithms, as well as the computer programs that have been developed to deal with the problem of identifying the minimal cut sets.

As an example, consider the fault tree for unavailability of water at the required pressure for sprinklers in a building (Figure 6.4).There are two separate water mains, main 1 and main 2. To attain the required pressure the water must be pumped. There is an electric pump and a backup diesel pump but they are not shown explicitly. The top event is "no water". The event "pumps not working" is denoted by x_1. The event "no water from main 1" is denoted by x_2 and the event "no water from main 2" is denoted by x_3.

Here, it is easy to see by inspection that there are two minimal cut sets: $\{x_1\}$ and $\{x_2, x_3\}$.

In every fault tree there are also *minimum path sets*. A path set is a set of basic events whose simultaneous non-occurrence ensures the non-occurrence of the top event. For the fault tree in Figure 6.4, it is easy to see by inspection that the path sets are $\{x_1, x_2\}$ and $\{x_1, x_3\}$.

6.4.5 Dual fault trees

To every fault tree there corresponds a dual fault tree obtained by replacing each event by its dual (i.e. its opposite), replacing OR gates by AND gates and

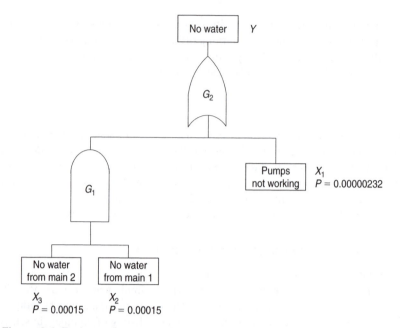

Figure 6.4 Fault tree for unavailability of water at the required pressure.

AND gates by OR gates. It should be noted that if the top event was originally a failure, the new top event is now the non-occurrence of the failure. For example, in the second fault tree of Figure 6.3, the dual of the top event "motor does not start" will be "motor starts".

The minimum cut sets for the dual fault tree are the minimum path sets for the original fault tree and vice versa.

6.4.6 Probability evaluation of fault trees

We aim to calculate the probability of occurrence of the top event from the probability of the basic events.

We define a *Boolean indicator function* for the top event as follows:

Let there be n basic events. Define, for the ith basic event

$$Y_i = \begin{cases} 1 & \text{if the event occurs,} \\ 0 & \text{otherwise.} \end{cases} \tag{6.1}$$

(Y_i is the *indicator function* of the ith basic event).

It is important to notice that for a Boolean variable Y, $Y^2 = Y$. It follows that, for any positive integer n,

$$Y^n = Y. \tag{6.2}$$

Let $\mathbf{Y} = (Y_1, Y_2, \ldots, Y_n)$ be the vector of indicator functions of the basic events. Define a function $\Psi(\mathbf{Y})$ by

$$\Psi(\mathbf{Y}) = \begin{cases} 1 & \text{if the top event occurs,} \\ 0 & \text{otherwise.} \end{cases} \tag{6.3}$$

The function Ψ is called the Boolean indicator function for the top event.

We now note that Boolean gates can be represented as follows.

Let $(Y_1, Y_2, \ldots, Y_m)$ be the indicator functions of the inputs of an AND gate. Then the indicator function Y of the output of the gate is equal to

$$Y = Y_1 Y_2 \cdots Y_m. \tag{6.4}$$

It is easy to see that $Y = 1$ if and only if all Y_i are equal to one.

Furthermore, let $(Y_1, Y_2, \ldots, Y_m)$ be the indicator functions of the inputs of an OR gate. Then the indicator function Y of the output of the gate is equal to

$$Y = 1 - (1 - Y_1)(1 - Y_2) \cdots (1 - Y_m), \tag{6.5}$$

for here $Y = 1$ if at least one of the Y_i is equal to one.

The function Ψ can be constructed from the fault tree by successively replacing the outputs of the gates by the appropriate functions of the inputs.

Example   Consider the water pressure fault tree in Figure 6.4. The output of the OR gate is denoted by G_2 and the output of the AND gate by G_1. Let the indicator functions of the events x_1, x_2 and x_3 be denoted by Y_1, Y_2 and Y_3, respectively. We then have

$$Y = G_2 \tag{6.6}$$

$$= 1 - (1 - G_1)(1 - Y_1) \tag{6.7}$$

$$= 1 - (1 - Y_2 Y_3)(1 - Y_1). \tag{6.8}$$

Thus, for the water pressure fault tree, $\Psi = 1 - (1 - Y_2 Y_3)(1 - Y_1)$.

Probability of top event

Let us note that, for any indicator function Y,

$$E(Y) = 1 \times P(Y = 1) + 0 \times P(Y = 0) \tag{6.9}$$

so that

$$E(Y) = P(Y = 1). \tag{6.10}$$

Similarly, the probability of the top event is given by $E[\Psi(\mathbf{Y})]$.

Boolean reduction

Let us assume for the moment that all basic events are statistically independent. We can then compute in principle the exact probability of the top event, using two basic properties of the expectation operator, namely:

1. $$E(Y_i + Y_j) = E(Y_i) + E(Y_j) \text{ always.} \qquad (6.11)$$

2. $$E(Y_iY_j) = E(Y_i)E(Y_j) \text{ if } Y_i \text{ and } Y_j \text{ are independent.} \qquad (6.12)$$

We expand the Boolean indicator function as a polynomial in the Y_i and get rid of the powers, using (6.2). We can then obtain $E(\Psi)$ by replacing the expectations of the basic event indicator variables by the respective event probabilities.

Example   Consider the Boolean indicator function for the water pressure event tree derived in Example 1. Expanding the function we find

$$\Psi = Y_1 + Y_2Y_3 - Y_1Y_2Y_3, \qquad (6.13)$$

from which we find that

$$E(\Psi) = E(Y_1) + E(Y_2)E(Y_3) - E(Y_1)E(Y_2)E(Y_3). \qquad (6.14)$$

This event tree is used in Ref. [66]. The probabilities of the three basic events are given as:

$$E(Y_1) = 0.0000232 \qquad (6.15)$$

$$E(Y_2) = 0.00015 \qquad (6.16)$$

$$E(Y_3) = 0.00015 \qquad (6.17)$$

and the three events are assumed independent.

Replacing the $E(Y_i)$ in (6.14) by their values, we find

$$E(\Psi) = 0.00000235. \qquad (6.18)$$

Approximate calculation of risk

Usually the probabilities of the basic events are small. It follows that the contributions to the probability of the top event by product terms are often negligible compared with the contributions of the terms that include the indicator function of just one basic event. Thus in (6.14), the first term $E(Y_1)$ dominates the other terms and we can conclude that $E(\Psi) \sim E(Y_1) = 0.0000232$.

Dependent events

If the occurrences of the basic events are not statistically independent, then the method just outlined is no longer valid. For example, components may be subject to a common environment, or they may share a load, so that failure of one component results in an increased load on other components. Methods have been developed to deal with this situation. The reader is referred to Ref. [36] for details.

It is often easiest to estimate the probability of the top event by Monte Carlo simulation, and this facility is usually available in most computer programs for fault tree analysis.

Time to occurrence of the top event

Let us assume that once a basic event has occurred it is not rectified. Let $T_1, T_2, \ldots, T_n$ be the times of occurrence of the basic events. Let

$$Y_i(t) = \begin{cases} 1 & \text{if } T_i \leq t \\ 0 & \text{otherwise.} \end{cases} \tag{6.19}$$

Then clearly $P(T_i \leq t) = E[Y_i(t)]$. Let us denote it by $F_i(t)$. Similarly, the probability that the top event occurs before t, which we denote by $F(t)$, is given by $E\{\Psi[\mathbf{Y}(t)]\}$. Thus $F(t)$ can be calculated by using Boolean reduction as before, with the $E(Y_i)$ replaced by $F_i(t)$.

Methods have been developed to tackle the case where components can be repaired or replaced, but the formulae are far more complicated.

Advantages and disadvantages of fault tree analysis

Disadvantages

1. There is a possibility of oversight and omission of significant failure modes.
2. It is difficult to apply Boolean logic to describe failures of system components that can be partially successful in operation and thereby affect the operation of the system, e.g. leakage through a valve.
3. For the quantitative analysis there is usually a lack of pertinent failure data. Even when there are data they may have been obtained from a different environment.

Advantages

1. It provides a systematic procedure for identifying faults that can exist within a system.
2. It forces the analyst to understand the system thoroughly.

7

Performance-based optimal design

7.1 Historical situation

The control mechanisms and organizational arrangements used at present in technologically advanced countries to respond to the threat of fires in buildings have obviously been largely successful, resulting in a very low death rate and comparatively low financial losses from building fires.

Historically, prescriptive building regulations have been an important component in design for fire safety in buildings. It is acknowledged that prescriptive design has resulted in the achievement of safety levels which the community appears to accept. Accordingly, it is appropriate to require that in the development of alternative designs for building fire safety and protection existing levels of protection should be maintained. However, that design approach does not generally result in the most cost-effective design solutions, nor in designs that maintain a consistent level of safety.

7.2 Performance-based design

In recent years, there has been a shift in design procedures for fire safety towards performance-based design. Instead of conforming to building regulations prescriptions, the design attempts to achieve a satisfactory performance of the fire fighting system. See for example Ref. [50]. Performance-based design has achieved wide acceptance as an alternative approach to prescriptive regulation in recent years.

The major objective of design for the effect of fire in buildings is to achieve satisfactory levels of life safety for:

1. Occupants of the building of fire origin
2. Occupants of adjoining buildings
3. Fire brigade personnel.

The level of fire safety in buildings is a reflection of the complex interaction between fire growth and spread and human behaviour. This depends on many features of the building including active and passive protection facilities, provisions for egress, occupant mobility and familiarity with the building, and building management. Thus, fire safety in buildings is a system consisting of many interacting subsystems.

Alternative building designs must ensure satisfactory levels of life safety. It would not be appropriate to subject the level of property protection in buildings to community regulation, but it is clearly the duty of the design engineer to offer to the owners and their insurers the design that will be most cost efficient while retaining the required level of life safety. There have been attempts to quantify the value of life in order to include it in the cost of the design. See for example Ramachandran [54]. But they have not achieved general acceptance. Thus the aim of performance-based design is two pronged: choosing the most cost efficient design while retaining an acceptable level of safety.

7.3 Risk assessment models

As mentioned in the Introduction, risk assessment models are needed to identify those combinations of building subsystems which provide the requisite level of safety in a cost-effective manner. Deterministic fire engineering design methods cannot be used for that purpose, because it is necessary to estimate the likelihood of all the possible fire scenarios in order to evaluate the likely cost and the likely safety level.

Risk assessment is defined as the process of assigning magnitudes and probabilities to the adverse effects resulting from fire in a building. The stimulus for adopting risk assessment as a fundamental component of decision making for managing a specific hazard is the recognition that:

1. the cost of eliminating all of the safety and environmental impacts from fires may be impossibly high,
2. design decision must be made on the basis of incomplete information.

Risk assessment provides rational criteria for the choice of remedial actions, including explicit consideration of uncertainty. It is obviously the preferred base for decision making.

7.4 Knowledge uncertainty and stochastic uncertainty

There are two types of uncertainties in design:

1. Knowledge uncertainty, which is due to lack of fundamental knowledge about the objects and the phenomena involved in the design. For example:
    (a) lack of knowledge about the amount and type of combustible materials that will be present in a room when the fire starts,
    (b) uncertainty about the accuracy of the fire modelling used,
    (c) uncertainty about the acceptable heat dose on a person.
2. Stochastic uncertainty, which is due to the intrinsic variability of the phenomena involved in the design. For example, the fire growth rate over a class of buildings.

Both types of uncertainty can be represented by probability distributions. These can be measured through statistical estimation on the basis of frequency distributions in suitable samples.

7.5 Expected value decisions

A common method used in many fields to account for uncertain costs is to weigh the possible outcomes by the probability of their occurrences. This is equivalent to the use of the *expected cost* as a summary measure. It can be demonstrated that when the values of the outcomes are properly expressed, this expected value is the only logical basis for choosing among alternative designs. For a full discussion of this topic see Benjamin and Cornell [39], Chapter 5.

It is customary in building design to specify a *design horizon* that depends on the expected useful life of the building, e.g. 20 years. Let C be the cost of fire fighting measures for a particular building. Let C_F be the expected fire loss within the design horizon. Finally let N be the expected number of deaths within the design horizon and let N_0 be the socially acceptable expected number of deaths within the design horizon for that particular building. Optimal fire fighting design then requires that the total expected fire loss C_T, given by

$$C_T = C + C_F \tag{7.1}$$

be minimized, subject to the level of safety condition

$$N \leq N_0. \tag{7.2}$$

Given a particular cost of fire fighting measures C, the design that minimizes the second term of equation (7.1), C_F, must be chosen. As expenditure on fire fighting measures C is increased, it is logical to expect the expected fire loss C_F to decrease. This indicates that there will often be an optimal value of C for which C_T is a minimum. This is illustrated in Figure 7.1. Of course the chosen

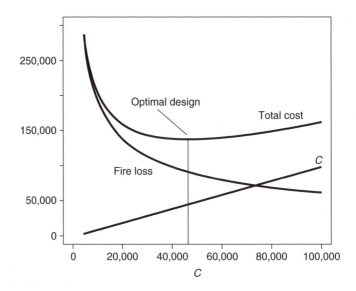

Figure 7.1 Optimal design for fire loss.

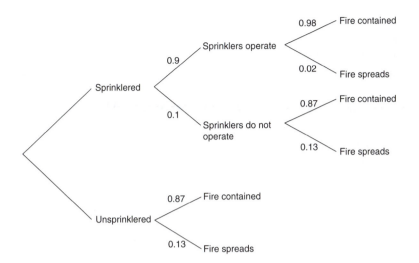

Figure 7.2 Event tree for sprinkler analysis.

design must ensure that the level of safety condition is satisfied. If it is not, then the expenditure on fire fighting measures must be increased until it is.

Example   The following example is based on a report by Rutstein and Cooke dated 1979 [56] and on Ramachandran [54].

A cost–benefit analysis is carried out to determine whether it is cost efficient to install sprinklers in a department store. The event tree considered is shown in Figure 7.2. It is assumed that the level of safety requirement will be satisfied whether or not sprinklers are installed.

We consider a department store of total floor area 12,000 m² consisting of four compartments of 3000 m² each. Following Rutstein [56], it is estimated that if a fire is confined to a compartment of area A m² the expected damaged area D m² will be given by the formula

$$D = CA^\beta. \tag{7.3}$$

For commercial buildings, $C = 0.95$. If there are no sprinklers, $\beta = 0.5$ and if there are sprinklers, $\beta = 0.25$. Thus, for our 3000 m² compartment the expected damaged area will be $0.95 \times 3000^{0.5} = 52$ m² if there are no sprinklers. According to the UK fire statistics for 1984–1986, the probability of a fire spreading beyond an unsprinklered compartment is 0.13. In that case the expected damaged area can be assumed to be 4000 m². Thus the expected damaged area in an unsprinklered building will be

$$(52 \times (1 - 0.13)) + (4000 \times 0.13) = 565 \, \text{m}^2. \tag{7.4}$$

On the other hand, for a compartment with sprinklers, if a fire is confined to the compartment, the expected damaged area is $0.95 \times 3000^{0.25} = 7$ m², provided the sprinklers operate, and the probability of a fire spreading beyond the compartment is 0.02. If this happens, the expected damaged area will be assumed to be 4000 m^2 as for the unsprinklered case. Thus the expected damaged area will be, conditional on the sprinklers operating,

$$(7 \times (1 - 0.02)) + (4000 \times 0.02) = 87 \, \text{m}^2. \tag{7.5}$$

The probability of the sprinklers not operating in a fire is estimated to be 0.1. In that case the expected damaged area will be taken to be equal to the damaged area for an unsprinklered compartment. So the unconditional expected damaged area for a sprinklered compartment will be

$$(87 \times (1 - 0.1)) + (565 \times 0.1) = 135 \, \text{m}^2. \tag{7.6}$$

According to Rutstein [56], the annual probability p of a fire starting in an area $A \, \text{m}^2$ is given by the formula

$$p = K A^\alpha \tag{7.7}$$

where $K = 0.000066$ and $\alpha = 1.0$ for department stores. Thus for our example $p = 0.000066 \times 12{,}000 \approx 0.8$. So the expected annual saving in damaged area due to sprinklers will be

$$(565 - 135) \times 0.8 = 344 \, \text{m}^2. \tag{7.8}$$

Hence, at a loss rate of £480/m^2 (at 1979 prices), the annual saving in financial terms is likely to be £165,000. Assuming a design horizon of 20 years, the total saving will be £3,300,000.

The cost of installing sprinklers (at 1979 prices) can be assumed to be £17/m^2. Thus the cost of installing sprinklers in the department store will be £204,000. Thus there can be no question in this case that the installation of sprinklers is justified.

It must however be pointed out that most property owners do not actually optimize their fire fighting expenditure on the basis of a calculated expected fire loss. They rely on fire insurance to cover their actual fire loss. In many countries, e.g. the UK, insurance companies offer a significant discount on their premiums for installation of sprinklers. In that case, the cost of installing sprinklers should be weighed against the premium discount rather than the expected fire loss. In some countries, e.g. Australia, insurance companies do not offer discounts for installing sprinklers. But in situations where it would be sensible to install sprinklers they would refuse insurance cover unless sprinklers are installed. In that case, of course, there is no point in carrying out the kind of calculation just described.

8

Risk analysis of fire initiation

8.1 Ignition frequency

For quantitative estimation of fire risks, reliable ignition frequency derived from fire statistics is a prerequisite. Annual ignition frequency obviously depends on the type of building. The average annual ignition frequency is the probability of a building catching fire per unit of floor area (measured in square metres) per annum. Average ignition frequency for three European locations and different building categories is given in Table 8.1

8.1.1 Dependence on the floor area

It has been known for some time [52,53] that within each building category ignition frequency also depends on the floor area of the building. Let $f(A)$ represent the average annual probability of a fire starting in a building in the category under study with area A, per unit area. Ramachandran [53] argued that since probability of a fire starting depends on the amount and nature of the ignition sources, and since the number of these sources increases with the area of the building, $f(A)$ is primarily a function of the floor area. Ramachandran further showed that if the logarithm of the floor area has either a negative exponential distribution or a normal distribution, then $f(A)$ is a power law, i.e. $f(A) \propto A^{-\lambda}$. However, analysis of the statistics of floor area shows that they can have distributions of many functional forms (Tillander and Keski-Rahkonen [68]). A more flexible functional form was therefore used by them to model the dependence of $f(A)$ on A. It is a generalization of a model originally proposed by a French probabilist called Barrois in 1835. The formula they employed is

$$f(A) = c_1 A^r + c_2 A^s. \qquad (8.1)$$

Table 8.1 Average ignition frequency ($\times 10^{-6}$) per square metre per annum

	London 1996–1999	Switzerland 1986–1995	Finland 1996–2001
Industrial and storage	6.9		11.1
Shop and commercial	22.0		6.6
Office	5.3		2.5
Dwellings		33.3	6.3
Public office		10.6	4.7
Industrial		11.6	9.6
All		32.3	9.0

Table 8.2 Parameters of the generalized Barrois model

Building category	c_1	$c_2 \times 10^6$	r	s	Number of fires
Residential buildings	0.010	5	−1.83	−0.05	4361
Commercial buildings	7×10^{-5}	6	−0.65	−0.05	356
Office buildings	0.056	3	−2.00	−0.05	140
Transport, fire-fighting and rescue-service buildings	7×10^{-5}	1	−0.65	−0.05	123
Institutional care buildings	2×10^{-4}	5	−0.61	−0.05	197
Assembly buildings	0.003	2	−1.14	−0.05	112
Educational buildings	0.03	3	−1.26	−0.05	122
Industrial buildings	3×10^{-4}	5	−0.61	−0.05	1038
Warehouses	3.82	2	−2.08	−0.05	405
Other buildings	1.18	100	−1.87	−0.20	2650

This generalized Barrois model was fitted to the observations on 13 different categories of buildings in Finland. The observations were from the years 1996–99. The parameters of the Barrois model are presented in Table 8.2.

The generalized Barrois model is useful in determining the ignition frequency of buildings with a floor area between 100 and 20,000 m^2.

8.1.2 Time distribution of ignition frequency

There are three cycles of time variation of ignition frequency: annual, weekly and hourly. The number of fires was observed to be highest during the cold winter months. Weekday variation was rather small. On Saturday values were 20% above average in residential buildings but significantly lower than average for industrial buildings and warehouses, because of low industrial activity. The number of building fires in all categories was lower during nighttime and higher during daytime.

8.2 Fire statistics

In most countries, details of fires in buildings are recorded whenever the fire brigade is called to a fire. Consequently there is much information available about such events. These statistics provide a useful resource that can help us understand the causes and consequences of fires. In this chapter, the fire statistics analysed are derived from the US National Fire Incident Reporting System (NFIRS) [3].

Basic data collected includes the number of casualties (civilian and fire fighter, injuries and fatalities) and many observations of the causes and other consequences of the fires. There are two types of records in the fire statistics: those related to the circumstances of the fire itself and those related to the personal characteristics of the casualties. The most complete record of fire-related factors is the Fire Incident data set (Form NFIRS-1). Personal characteristics of civilian casualties appear only in the Civilian Casualty Report (Form NFIRS-2). Unfortunately, the fire statistics do not provide any data on the severity of the injuries. Some of the injuries reported may well be due to health problems such as heart attacks or may have been sustained while escaping.

The analysis of fire statistics described in this chapter and the next chapter is based on Hasofer and Thomas [33].

8.3 Distribution of fire losses

In 1977, Rogers published an article [55] in which he studied the probability distribution of fire losses in the UK for various industrial occupancies. The purpose of the article was to evaluate the effect of sprinkler protection on fire losses. The data he analysed were yearly values over the period 1966–1972. Only individual fires for which the total damage to structure and content was £10,000 or more were included.

The purpose of the present section is to revisit Rogers' findings about the statistical properties of the data, using the far more extensive data available from the USA, as well as some more refined statistical techniques that have been recently developed.

Rogers assumed that the fire loss was lognormally distributed. Although he claimed that, based on previous work, this was a "reasonable" assumption, he was not able to justify it on the basis of the data presented, because the data available to him represented "only a small percentage of the number of fires", namely the upper tail, with the bulk of the distribution absent. The data presented in this section is obtained from NFIRS and does cover the full range of losses. The analysis will be given for one particular data set, namely hotels and motels for the year 1988.

The first task is to test whether the data are consistent with a lognormal distribution. We have from the outset a problem, since the probability of the value zero in a lognormal distribution is zero. In the data, however, there is a comparatively large proportion of zeroes. For 1988, the total number of fires listed was 3377 of which 943, or 27.9%, had a zero fire loss.

There were clearly two ways to deal with the problem. The first (and most obvious one) was to imagine that the zero values were actually distributed lognormally in the positive neighbourhood of the origin. The second one was to just disregard the zero values and fit a lognormal distribution to the loss values greater than zero. The fitting was carried out using a quantile–quantile plot (Section 3.10.5) of the logarithm to the base 10 of the non-zero fire loss (in US dollars) against the quantiles of the standard normal distribution. Figure 8.1 shows the quantile–quantile plot when the zeroes are taken into account, while Figure 8.2 shows the plot when the zeroes are ignored. It is quite clear that the second plot is a better fit. Numerically, this can be confirmed by calculating the correlation between the quantiles. For Figure 8.1 the correlation is 0.984, while for Figure 8.2 it is 0.994.

It might have been thought that it is quite possible that there were many more fires with negligible losses than shown in the data, but they were not reported. This would suggest that the fit would be improved by increasing the proportion of zeroes in the data. This, however, is not the case. Increasing the number

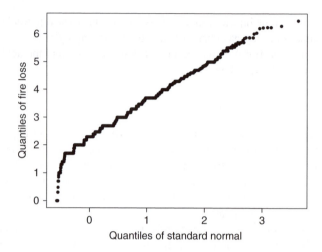

Figure 8.1 Lognormality of fire loss, including zeroes.

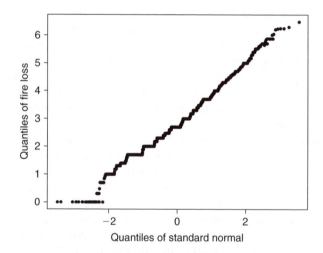

Figure 8.2 Lognormality of fire loss, excluding zeroes.

of zeroes actually decreases the correlation. For example, if the proportion of zeroes is increased to 50% the correlation decreases to 0.976.

The mean of the logarithm (to the base 10) of the fire loss was 2.85 and the standard deviation 1.02.

For full details of this topic see Ref. [31].

8.4 Fire factors affecting ignition

In attempting to design buildings for fire safety, whether in the context of design of individual buildings (as might be done by a fire safety engineer) or in designing building regulations that are intended to cover broad classes of buildings (as might be done by building regulatory bodies), it is a basic requirement that the designers have an appreciation of the circumstances in which unwanted events (such as injury and death) occur and the frequencies at which they occur. It is obvious that there is little value in concentrating on events that do not occur or that occur very infrequently, particularly when it is apparent that the vast majority of losses occur through relatively frequent events. This is not to say that infrequent events should be ignored, merely that they should be kept in perspective. Indeed, when the potential consequences of a particular infrequent event are unacceptably severe, it is commonsensical to adopt specially designed measures to minimize the risk.

Casualties (injuries and fatalities) occur relatively infrequently in fires that are notified to fire brigades. Also, most reported fires result in very few casualties but nevertheless the great majority of casualties occur in these fires.

Injuries and fatalities that occur as a consequence of fires in buildings may be viewed from several perspectives. Most injuries and fatalities are the result of relatively frequent occurrences – particular types of fires and situations that usually result in no casualties but occasionally in one or two injuries and/or fatalities. Thus it is important to recognize and address in design fire locations and types that cumulatively result in large numbers of fires with casualties.

Additional measures that might indicate fire types or locations that warrant particular attention are the total number of fatalities, the proportion of fatalities among casualties and the average number of fatalities or casualties per fire.

There are many ways of assessing the data. Take civilian fatalities as an example. We can look at civilian fatalities in many different ways.

The first and most obvious way is simply to look at the total number of fatalities. If we divide the data up into factor categories, for example, categories that classify the fire simply by where it started (what type of room) then we can look at how many of the fatalities occurred in fires associated with each room. Experience has shown that some rooms will have far more fatalities associated with them than others. Thus, if our objective is to reduce fatalities in fires, it would be important to look at the rooms with the most fatalities first. So simply the number of fatalities in each category is important.

But if we are trying to make comparisons between categories, one of the things we are likely to find is that the number of fires in each category is different as well as the number of fatalities. Thus in comparing categories we have to balance the number of fires in that category with the number of fatalities in the category. Thus the proportion of fires that result in fatalities is another measure that might be important. Some categories have a far higher proportion of fires with fatalities than others.

Similarly, given that a fatality occurs, another way of comparing categories is to look at the average number of fatalities that occurs in each fire. We might consider it important to concentrate on fires that consistently result in multiple fatalities compared with those that result in just one fatality.

Another way of viewing fires might be to consider not simply fatalities but all casualties. In this case we might wish to consider the total number of casualties.

Table 8.3 Frequency distribution of casualties per fire (apartment fires, USA, 1993)

Number of casualties	Frequency
0	42,470
1	1695
2	362
3	115
4	38
5	36
6	22
>6	26

Alternatively we might consider the proportion of casualties that result in a fatality.

The main purpose of this section is to identify and quantify the effect of fire-related factors that significantly affect the danger to occupants in fires. The measures of danger to occupants studied are:

1. Number of casualties per fire
2. Proportion of fatalities among casualties
3. Proportion of fires with casualties
4. Number of fatalities per fire.

The analysis was carried out on a restricted statistical database, namely apartment fires (code 42: apartments, tenements, flats) during the year 1993.

The data set used was the Fire Incident data (Form NFIRS-1), in which each row represents a fire. There were 44,764 fires in the Fire Incident data of which 2294 resulted in at least one casualty, with a total of 3463 casualties.

8.4.1 Distribution of the number of casualties per fire

In the Fire Incident data, the total number of casualties per incident (i.e. per fire) was calculated by adding the number of non-fatal injuries to the number of fatalities.

Table 8.3 shows the frequency distribution of the number of casualties per fire.

Clearly the great majority of reported fires (about 95%) do not cause any casualties. To further analyse the frequency distribution of the number of casualties per fire, we concentrate on the fires that have caused at least one casualty. It then turns out that that frequency distribution can be reasonably approximated by a *displaced Poisson distribution*. By this is meant that if X is the random variable that represents the number of casualties per fire, then the probability function of X, given that X is greater than zero, is such that $X - 1$ has approximately a Poisson distribution. For the 1993 data the appropriate parameter λ is about 0.35. This is illustrated in Figure 8.3.

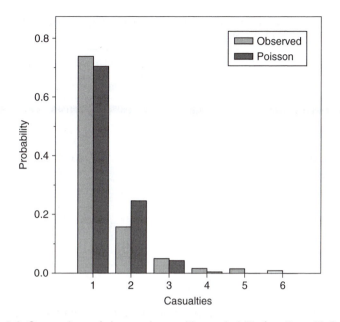

Figure 8.3 Comparison of observed casualties probability function with Poisson probability function.

8.4.2 Fire factors affecting the number of casualties per fire

A regression (Section 3.14) of the number of casualties per fire was carried out, for all 2294 fires resulting in at least one casualty, on a variety of factors, thought to be relevant. They were:

1. Extent of fire damage
2. Area of fire origin (i.e. the room or space in which the fire originated)
3. Level of fire origin
4. Number of storeys
5. Equipment involved in ignition
6. Form of heat of ignition
7. Type of material ignited
8. Form of material ignited
9. Ignition factor
10. Construction type
11. Extent of fire damage
12. Detector performance
13. Sprinkler performance.

The most important factor turned out to be the area of fire origin, followed by the extent of fire damage, the type of material ignited and the ignition factor.

The level of fire origin, the number of storeys, the construction type, detector performance and sprinkler performance were the least significant factors. For a full discussion of this (perhaps unexpected) result see Thomas [64].

Table 8.4 Significant areas of fire origin causing a large number of casualties (apartment fires, USA, 1993)

Key	Mean casualties per fire	Number of fires with casualties	Total Number of fires	Percentage of fires with casualties
Hallway, corridor, mall	1.82	34	1048	3.24
Lounge area	1.69	314	2988	10.51
Sleeping room for under five persons	1.62	614	5957	10.31
Kitchen, cooking area	1.28	885	20,862	4.24
Lavatory, locker room, cloakroom	1.62	37	967	3.83
Product storage room or area	1.73	11	97	11.34
Supply storage room or area	2.00	15	348	4.31
Crawl space, substructure space	2.21	19	461	4.12
Structural areas not otherwise classified	2.29	14	169	8.28

Identification of the most important areas of fire origin that result in many casualties

The Fire Incident data code the areas of fire origin according to 45 different categories. Many of these have a very small representation in the data and many others cause very few casualties. Table 8.4 lists the identified areas of fire origin that resulted in a higher than average number of casualties (1.51), as well as those that have a significant representation in the data. The areas where most fires with casualties occurred (many more than in the other areas listed) were the kitchen (i.e. the cooking area), the lounge area and sleeping rooms for under five persons.

Identification of the most important types of material ignited that result in many casualties

The Fire Incident data code the types of material ignited according to 66 different categories. Many of these have a very small representation in the data and many others cause very few casualties. Table 8.5 lists the identified types of material ignited that tend to cause a higher than average number of casualties (1.51), as well as those that have a significant representation in the data. The items causing most fires with casualties (many more than the other listed) were fabrics, whether man-made or cotton and finished goods.

Identification of the most important ignition factor categories that result in many casualties

The Fire Incident data code the Ignition Factor of individual fires according to 47 different categories. Many of these have a very small representation in the data and many others cause very few casualties. Table 8.6 lists the categories

Table 8.5 Significant types of material ignited causing a large number of casualties (apartment fires, USA, 1993)

Key	Mean casualties per fire	Number of fires with casualties	Total number of fires	Percentage of fires with casualties
Gasoline	1.71	28	347	8.07
Fat, grease (food) (included are butter, tallow margarine and lard)	1.21	453	9673	4.68
Plastic; insufficient information to classify further	1.53	55	1326	4.15
Polyurethane (including polyisocyanurates)	1.58	24	278	8.63
Rubber (including synthetic rubber)	2.00	27	675	4.00
Food, starch (excluded are fat and grease)	1.25	113	4279	2.64
Sawn wood	1.70	97	3094	3.14
Paper, untreated, uncoated. (excluded are waxed papers)	1.37	67	2525	2.65
Fabric, textile, fur; insufficient information to classify further	1.28	79	613	12.89
Man-made fabric, fibre, finished goods (excluding rayon)	1.63	234	2577	9.08
Cotton, rayon, cotton fabric, finished goods (included are canvases and all polyester cotton blends)	1.51	402	4326	9.29
Multiple fires started with more than one type of material	1.96	72	1049	6.86

of ignition factor that tend to cause a higher than average number of casualties (1.51), as well as those that have a significant representation in the data. The items causing most fires with casualties (many more than the other listed) were misuse of heat of ignition by children and incendiary (not during civil disturbance).

8.4.3 The proportion of fatalities among casualties

In this section the proportion of casualties that were fatalities in the fire incidents where there was at least one casualty was regressed against the factors used in the previous section.

The important factors turned out to be similar to those for the number of casualties, namely, the area of fire origin, the extent of fire damage, the type of material ignited and the ignition factor.

Similarly, the least significant factors were the level of fire origin, the number of storeys, the construction type, detector performance and sprinkler performance.

Table 8.6 Significant ignition factor categories causing a large number of casualties (apartment fires, USA, 1993)

Key	Mean casualties per fire	Number of fires with casualties	Total number of fires	Percentage of fires with casualties
Incendiary, not during civil disturbance	1.77	122	3164	3.86
Suspicious, not during civil disturbance	1.63	89	2161	4.12
Misuse of heat of ignition (no details available)	1.58	31	378	8.20
Abandoned, discarded material (included are discarded cigarettes, cigars and the like)	1.45	284	3785	7.50
Falling asleep	1.32	165	1497	11.02
Inadequate control of open fire	1.45	11	471	2.34
Misuse of heat of ignition by children	1.85	210	2051	10.24
Misuse of heat of ignition not classified	1.21	58	1077	5.39
Misuse of material ignited (no details available)	1.50	14	133	10.53
Combustible too close to heat	1.42	108	2012	5.37
Misuse of ignited material by children	1.91	33	322	10.2
Short circuit, ground fault	1.49	111	2697	4.12
Electrical failure (excluding short circuit, ground fault)	1.90	42	906	4.64
Lack of maintenance, worn out	1.45	11	1049	1.05

Identification of the most important areas of fire origin that result in a high proportion of fatalities

Table 8.7 lists the identified areas of fire origin that tend to cause a high proportion of fatalities, as well as those that have a significant representation in the data. The column headed "mean proportion of fatalities" represents the ratio of the total number of fatalities to the total number of casualties in each category. The areas where most fires with casualties occur are, as pointed out in Section 8.4.2, the lounge area and sleeping rooms for under five persons. In addition, however, the kitchen or cooking area now appears as an area where a very large number of fires with casualties occurs. The mean proportion of fatalities is comparatively high even though the mean number of casualties is low (1.28).

Table 8.7 Significant areas of fire origin causing a high proportion of fatalities (apartment fires, USA, 1993)

Key	Mean proportion of fatalities	Number of fires with casualties	Total number of fires	Percentage of fires with casualties
Hallway, corridor, mall	0.11	34	1048	3.2
Lounge area	0.149	314	2988	10.5
Sleeping room for under five persons	0.077	614	5957	10.3
Kitchen, cooking area	0.028	885	20,862	4.2
Lavatory, locker room, cloakroom	0.05	37	967	3.8
Closet	0.059	50	805	6.2
Supply storage room or area	0.033	15	348	4.3
Heating equipment room or area	0.06	25	957	2.6
Crawl space, substructure space	0.024	19	461	4.1
Exterior balcony, open porch	0.118	14	649	2.18

Identification of the most important types of material ignited that result in a high proportion of fatalities

Table 8.8 lists the identified types of material ignited that tend to cause a high proportion of fatalities, as well as those that have a significant representation in the data. The most notable are paper and fabrics.

Identification of the most important ignition factors that result in a high proportion of fatalities

Table 8.9 lists the identified ignition factors that tend to cause a high proportion of fatalities, as well as those that have a significant representation in the data. The main new item is "abandoned, discarded material (including cigarettes, cigars and the like)". It causes a large number of fires with a high mean proportion of fatalities, even though the mean number of casualties (1.45) is comparatively low.

Table 8.9 illustrates a major problem with fire data, in that ignition factors include heat sources (e.g. smoking materials), motives (e.g. incendiary) and impairment (e.g. alcohol). This is such a mixed bag that it is difficult to analyse the data meaningfully.

Influence of the extent of fire damage on the proportion of fatalities

Table 8.10 documents the influence of the extent of fire damage on the proportion of fatalities. The table shows that not only does the mean number of casualties rise with the extent of fire damage, but also the proportion of fatalities among casualties.

Table 8.8 Significant types of material ignited causing a high proportion of fatalities (apartment fires, USA, 1993)

Key	Mean proportion of fatalities	Number of fires with casualties	Total number of fires	Percentage of fires with casualties
Flammable, combustible liquid (not further classified)	0.148	18	215	8.37
Gasoline	0.083	28	347	8.07
Class IIIB combustible liquid. Flashpoint at or above 93.4°C	0.055	43	672	6.40
Fat, grease (food) (Included are butter, tallow margarine and lard)	0.013	453	9673	4.68
Plastic; insufficient information to classify further	0.048	55	1326	4.15
Food, starch (excluded are fat and grease)	0.014	113	4279	2.64
Sawn wood (included are all finished lumber)	0.048	97	3094	3.14
Paper, untreated, uncoated (excluded are waxed papers)	0.065	67	2525	2.65
Wood, paper (not in other classifications)	0.077	19	441	4.31
Fabric, textile, fur (not further classified)	0.079	79	613	12.89
Man-made fabric, fibre, finished goods (excluding rayon)	0.121	234	2577	9.08
Cotton, rayon, cotton fabric, finished goods	0.097	402	4326	9.29
Fabric, textile (not in other classifications)	0.075	18	248	7.26
Multiple fires started with more than one type of material	0.121	72	1049	6.86

8.4.4 The number of fatalities per fire in the Fire Incident data

Table 8.11 gives the frequency distribution of fatalities among the 2294 fires that had at least one casualty.

The great majority of the 2294 fires with casualties (about 91%) do not cause any fatalities. As for casualties, the frequency distribution of fires with at least one fatality can be reasonably approximated by a *displaced Poisson distribution* with parameter 0.2. The total number of fatalities was 213.

Factors affecting the number of fatalities per fire causing at least one casualty

Analysis showed that the three most important factors affecting the number of fatalities per fire are, in order of importance, the form of material ignited, the form of heat of ignition and the ignition factor.

Table 8.9 Significant types of ignition factors causing a high proportion of fatalities (apartment fires, USA, 1993)

Key	Mean proportion of fatalities	Number of fires with casualties	Total number of fires	Percentage of fires with casualties
Incendiary (not during civil disturbance)	0.088	122	3164	3.86
Misuse of heat of ignition (not further classified)	0.204	31	378	8.20
Abandoned, discarded material (including cigarettes, cigars and the like)	0.102	284	3785	7.50
Misuse of heat of ignition: children with, children playing	0.082	210	2051	10.24
Unconscious; mental, physical impairment; drug, alcohol stupor	0.071	34	198	17.17
Combustible too close to heat	0.085	108	2012	5.37
Misuse of material ignited: children with, children playing	0.032	33	322	10.25
Misuse of material ignited not classified above	0.047	30	364	8.24
Property too close to (included are exposure fires)	0.308	18	780	2.31
Operational deficiency (not further classified)	0.090	10	156	6.41
Accidentally turned on, not turned off	0.077	34	846	4.02
Unattended	0.020	400	11,040	3.62

Table 8.10 Influence of extent of fire damage on the proportion of fatalities (apartment fires, USA, 1993)

Key	Mean proportion of fatalities	Number of fires with casualties	Total number of fires	Percentage of fires with casualties
Confined to object of origin	0.016	516	20,390	2.53
Confined to part of room or area of origin	0.035	544	8334	6.53
Confined to room of origin	0.064	473	5893	8.03
Confined to fire-rated compartment of origin	0.082	108	746	14.48
Confined to floor of origin	0.106	191	1597	11.96
Confined to structure of origin	0.157	352	3267	10.77
Extended beyond structure of origin	0.176	49	537	9.12

Identification of the most important forms of material ignited that result in many fatalities

There are 26 categories represented in the fires that lead to fatalities. But only 3 categories tend to cause a comparatively high number of fatalities while having

Table 8.11 Frequency distribution of fatalities per fire (apartment fires, USA, 1993)

Number of fatalities	Frequency
0	2081
1	179
2	20
3	11
4	0
5	3

a significant representation in the data. They are:

1. Upholstered sofa, chair, vehicle seat
2. Bedding, blanket, sheet, comforter
3. Multiple forms of material first ignited.

Identification of the most important forms of heat of ignition that result in many fatalities

There are 26 categories represented in the fires that lead to fatalities. But only 4 categories tend to cause a comparatively high number of fatalities while having a significant representation in the data. They are:

1. Heat from properly operating electrical equipment
2. Match
3. Lighter (flame type)
4. Cigarette.

Identification of the most important ignition factors that result in many fatalities

There are 24 categories represented in the fires that lead to fatalities. But only 4 categories tend to cause a comparatively high number of fatalities while having a significant representation in the data. They are:

1. Incendiary, not during civil disturbance
2. Abandoned, discarded material
3. Misuse of heat of ignition: children with, children playing
4. Combustible too close to heat.

8.5 Conclusions

The fire data analysis presented allow us to determine the main fire factors that lead to high casualties and high fatalities in apartment fires, whether in absolute terms or relative to the number of casualties.

The number of casualties, as well as the proportion of fatalities both depend mainly on the same factors:

1. The extent of fire damage
2. The area of fire origin

3. The material ignited
4. The ignition factor.

The worst areas of fire origin are lounge areas, sleeping rooms and kitchens, while the most lethal materials are fabric and flammable liquids. The worst ignition factors are children playing with fire and cigarettes.

On the other hand, the absolute number of fatalities does not appear to depend significantly on the factors listed above, apart from the ignition factor. It depends mainly on two additional factors:

1. The form of material ignited
2. The form of heat of ignition.

As far as the absolute number of fatalities is concerned, the worst forms of material ignited are sofas, chairs and bedding. The worst forms of heat of ignition are electrical equipment, matches, lighters and cigarettes, and the worst ignition factors are children, "incendiary" and combustible materials too close to heat.

9

Personal factors in a fire

In this chapter the Civilian Casualties Report (Form NFIRS-2) for Apartment Fires (code 42: apartment, tenements, flats) during the year 1993 will be used to determine the most important personal factors affecting the proportion of fatalities among casualties in a fire (for a description of NFIRS see Section 8.2).

In addition to the factors relating to the fire that were studied in Chapter 8 using the Fire Incident data, the Civilian Casualty data allow the study of personal factors as well.

In the Civilian Casualty data, under the label "severity", it is indicated whether the casualty was just injured or killed. Thus the proportion of fatalities can be studied.

A study of the fire-related factors in the Civilian Casualty data does not add anything new. This is not surprising, since the data described in both files are identical, albeit couched in different terms. There are insignificant differences due to the gaps in the data not being identical.

The most important fire-related factors are the area of fire origin, the extent of fire damage, the type of material ignited and the ignition factor.

9.1 Personal factors affecting the proportion of fatalities

The personal factors considered in this chapter are as follows:

1. Sex
2. Familiarity with structure
3. Location at ignition
4. Condition before injury
5. Condition preventing escape
6. Activity at time of injury
7. Cause of injury
8. Age group.

The most important factors turned out to be, in order of importance:

1. Condition preventing escape
2. Condition before injury
3. Activity at time of injury
4. Location at ignition
5. Cause of injury.

Table 9.1 The influence of age group on the proportion of fatalities (apartment fires, USA, 1993)

Range	Mean proportion of fatalities	Number of casualties	Percentage of total casualties
0–4	0.10	393	15.82
5–9	0.13	97	3.90
10–19	0.07	192	7.73
20–39	0.06	1009	40.62
40–59	0.07	445	17.91
60–69	0.09	137	5.52
>70	0.06	211	8.49

Age group as a whole was not highly significant. This, however, does not contradict the fact, documented in Table 9.1, that the very young and the old are particularly susceptible to a high proportion of fatalities.

9.1.1 Influence of the most important conditions preventing escape that result in a high proportion of fatalities

The conditions preventing escape that tend to cause a high proportion of fatalities, while having a significant representation in the data are:

1. No time to escape (7.4%)
2. Fire between casualty and exit (9.3%)
3. Clothing on casualty burning (1.7%)
4. Incapacitated prior to ignition (5.0%).

The numbers between brackets are the percentage of total fatalities. The conditions preventing escape that cause the largest number of casualties are "fire between casualty and exit" and "no time to escape".

9.1.2 Influence of the most important conditions before injury that result in a high proportion of fatalities

The conditions before injury that tend to cause a high proportion of fatalities, while having a significant representation in the data are:

1. Asleep (27.7%)
2. Bedridden (1.4%)
3. Impaired by drugs, alcohol (3.1%)
4. Too old to act (3.5%).

The numbers between brackets are the percentage of total fatalities. The one condition that is responsible for by far the largest number of casualties (although not the highest proportion of fatalities) is "asleep".

9.1.3 Influence of the most important activities at time of injury that result in a high proportion of fatalities

The activities at time of injury that tend to cause a high proportion of fatalities, while having a significant representation in the data are:

1. Escaping (20.7%)
2. Sleeping (17.8%)
3. Unable to act (4.2%)
4. Irrational action (3.6%).

The numbers between brackets are the percentage of total fatalities.

9.1.4 Influence of the most important locations at ignition that result in a high proportion of fatalities

The locations at ignition that tend to cause a high proportion of fatalities, while having a significant representation in the data are:

1. Intimately involved with ignition (13.1%)
2. In room of fire origin (24.9%)
3. On floor of fire origin (23.1%)
4. In building of fire origin (21.0%).

The numbers between brackets are the percentage of total fatalities.

9.1.5 Influence of the most important causes of injury that result in a high proportion of fatalities

The causes of injury that tend to cause a high proportion of fatalities, while having a significant representation in the data are:

1. Trapped (2.6%)
2. Exposed to fire products: flame, smoke, gas, heat (78.5%)
3. Exposed to chemicals, radiation (1.0%).

The numbers between brackets are the percentage of total fatalities. Clearly, the overwhelming majority of casualties that resulted in a fatality was due to exposure to fire products.

9.1.6 The sex factor

The data analysed in this study do not show any significant difference between sexes as far as the proportion of fatalities is concerned. Indeed, the mean proportion of fatalities for males and females was 0.077 and 0.074, respectively, while the number of casualties was 1278 and 1197, respectively. This conclusion appears to contradict the conclusions of other studies. It may be a peculiarity of the apartment fires in 1993. Analysis of other fire data for other years will be needed before a final conclusion can be reached.

9.1.7 Age and gender

Table 9.1 gives an analysis of the influence of the age group on the proportion of fatalities. The analysis confirms the well-known fact that the age groups most susceptible to a high proportion of fatalities are the young, particularly the very young, and the old.

9.2 Conclusions

As far as personal factors are concerned, it is only possible, because of the structure of the data, to analyse the proportion of casualties that leads to a fatality.

The worst personal factors leading to a high proportion of fatalities are:

1. The condition preventing escape
2. The condition before injury
3. The activity at time of injury
4. The location at ignition
5. The cause of injury.

The worst conditions preventing escape were: burning clothes, incapacitation, fire between the casualty and exit, and no time to escape.

The worst conditions before injury were: bedridden, too old, drugs and alcohol, and asleep.

The worst activities at time of injury were: unable to act, sleeping, escaping and irrational behaviour.

The worst locations at ignition were: involved with ignition, in room of fire origin, in building of fire origin and on floor of fire origin.

Finally, by far the worst cause of injury leading to a fatality was exposure to fire products: flame, smoke, gas and heat.

10

Probabilistic modelling of barrier resistance

10.1 Introduction

The aim of the research described in this chapter is to determine the probability characteristics of the time to failure for light timber-framed walls with gypsum board exposed to a fire. The reader is referred to Refs [16,17,71] for further details.

The modes of failure considered are:

1. "Insulation, average temperature" mode: the average temperature rise on the ambient side of the wall exceeds some nominated value. In Australian Standard AS1530.4 [48] this value is 140°C.
2. "Insulation, maximum temperature" mode: the maximum temperature rise on the ambient side of the wall exceeds some nominated value. In Australian Standard AS1530.4 [48] this value is 180°C.
3. "Integrity initiated failure" mode: gypsum boards begin to crack in areas where the temperature rises above 400°C. Since this temperature exceeds insulation failure temperatures, failure due to integrity alone will never happen. However, pieces of gypsum on the fire side, formed by the cracking, will "slough off" (i.e. fall away) when the temperature exceeds 700°C. The fire then directly impinges on the timber studs and destroys them in a matter of minutes. Failure in this manner is referred to as "integrity initiated failure".
4. Structural collapse.

10.2 Overview of the time of failure model

Figure 10.1 shows the general flowchart for the time of failure model.
It is made up of three submodels:

1. The fire severity submodel
2. The heat transfer submodel
3. The structural response submodel.

Each of these three submodels will be described in subsequent sections.

10.3 The probability of failure model

The time of failure model was incorporated into a Monte Carlo analysis to yield the output probabilities of failure with time. The flowchart of the probability of failure model is given in Figure 10.2.

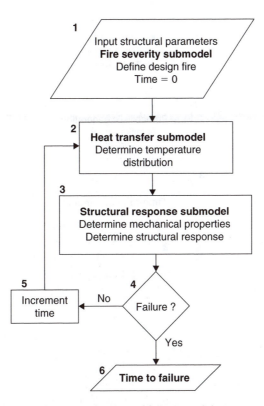

Figure 10.1 General flowchart for time of failure model.

The Monte Carlo analysis commences each simulation of the time of failure model with the probabilistic generation of random variables. As failures are encountered at the end of each simulation of the time of failure model, failure records are updated in the probability of failure model and further simulations are carried out until the number of simulations specified is completed.

All random variables are taken to be lognormal.

The thermal properties of wood and gypsum board, as well as the mechanical properties of timber, were taken to be temperature dependent. However, it was assumed that the temperature variation affected only the mean of the variables, leaving the coefficient of variation fixed. Thus, for every variable x a mean-independent lognormal random parameter α_x was generated with mean 1 and an appropriate coefficient of variation. The value of the random variable at temperature T was then calculated as

$$x = E_T(x)\alpha_x, \tag{10.1}$$

where $E_T(x)$ is the appropriate mean for the temperature T. The appropriate means are illustrated in Figures 10.3–10.9.

The mean conductivities of wood and gypsum board were assumed to be dependent on the random density ρ. This dependence was taken into account by generating first the random density ρ and then evaluating the mean

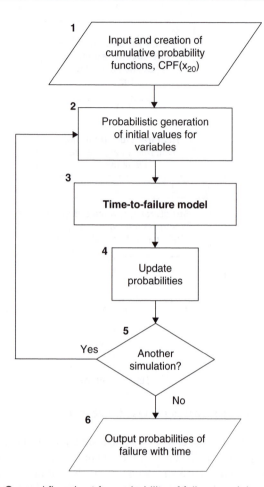

Figure 10.2 General flowchart for probability of failure model.

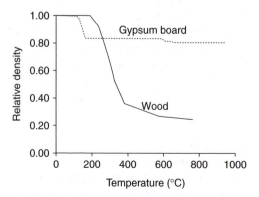

Figure 10.3 Relative densities of gypsum board and wood with temperature.

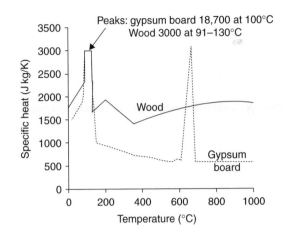

Figure 10.4 Specific heats of gypsum board and wood with temperature.

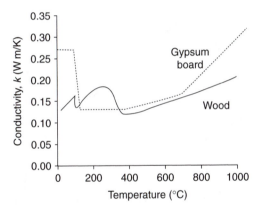

Figure 10.5 Conductivities of gypsum board and wood with temperature.

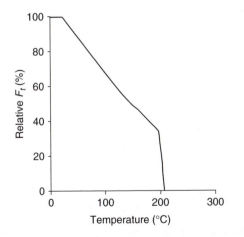

Figure 10.6 Relative tensile strength of timber versus temperature.

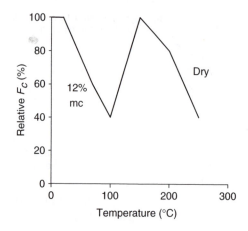

Figure 10.7 Relative compression strength of timber versus temperature.

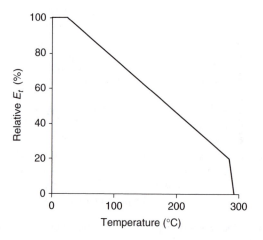

Figure 10.8 Relative elastic modulus of timber in tension versus temperature.

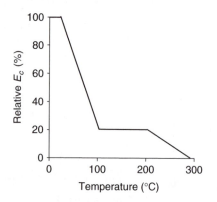

Figure 10.9 Relative elastic modulus of timber in compression versus temperature.

Table 10.1 Values adopted for random variables

Variable	Mean	Coefficient of variation	
		Standard fire	Real fire
Fire variable			
Fuel (kg/m^2)	10	NA	0.70
Thermal properties of wood			
Density, ρ (kg/m^3)	470 (at 20°C) Figure 10.3	0.06	0.06
Specific heat (J/kg K)	Figure 10.4	0.02	0.02
Conductivity (W/m K)	$(0.4 + 0.6\alpha_\rho) \times k(T)$ $k(T)$ as in Figure 10.5	0.02	0.02
Thermal properties of gypsum board			
Density, ρ (kg/m^3)	810 (at 20°C) Figure 10.3	0.02	0.02
Specific heat (J/kg K)	Figure 10.4	0.02	0.02
Conductivity (W/m K)	$(0.1 + 0.9\alpha_\rho) \times k(T)$ $k(T)$ in Figure 10.5	0.02	0.02
Mechanical properties of timber (MPa)			
Compression strength	24 (at 20°C) F_c in Figure 10.7	0.07	0.07
Tensile strength	24 (at 20°C) F_t in Figure 10.6	0.07	0.07
Elastic modulus	7400 (at 20°C) E in Figure 10.8 or 10.9	0.07	0.07
Structural loads (kN/stud)			
Dead load	5.67	0.10	0.10
Live load	2.33	0.70	0.70

conductivities as a function of the generated α_ρ and the temperature. Finally, the random conductivity lognormal parameters were generated, and the random conductivities evaluated, using equation (10.1).

The input variables that were chosen to be random, together with their coefficient of variation, are listed in Table 10.1.

10.4 The fire severity submodel

Two fire severity models were used:

1. The standard fire specified in Australian Standard AS1530.4.
2. The well-established fire severity model of Kawagoe and Lie [42].

10.4.1 The standard fire

In the standard fire (AS1530.4 [48]), temperature is controlled to vary with time, as closely as possible in accordance with the following relationship:

$$T_1 - T_0 = 345 \log_{10}(8t + 1), \tag{10.2}$$

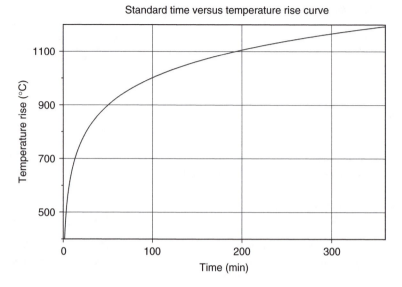

Figure 10.10 Standard fire rise curve.

where

T_1 = furnace temperature at time t, in degrees Celsius;
T_0 = initial furnace temperature, in degrees Celsius;
t = time into the test, measured from ignition, in minutes.

The relationship is illustrated in Figure 10.10.

10.4.2 The Kawagoe and Lie fire severity model

The model is based on the balance of all heat flows into an enclosure including:

1. Radiation lost through openings
2. Heat content of air flowing in through openings
3. Heat content of smoke flowing out through openings
4. Heat absorbed by boundaries
5. Heat absorbed by gases as the gases rise in temperature
6. Heat produced by combustion.

Let

$$F = \text{opening factor}$$

$$= \frac{A\sqrt{H}}{A_T},$$

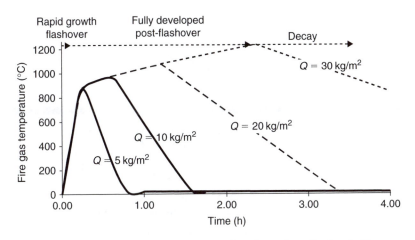

Figure 10.11 Fire gas temperature versus time for various fire loads Q.

where

A = area of openings (m^2);
A_T = total internal surface area of enclosure;
      with no deductions for openings (m^2);
H = average height of openings (m).

Let

T_f = fire gas temperature;
Q = fuel density in kilogram of wood per square metre of
      total internal surface area;
t = time in hours.

Examples of plots of T_f versus time for various values of Q are shown in Figure 10.11, with $F = 0.05$ m$^{0.5}$. The size of enclosure adopted was 3.00 m height, 5.00 m breadth and 4.00 m depth. The opening adopted was 1.20 m height by 2.00 m width.

The purpose of the use of the Kawagoe and Lie fire severity model was to demonstrate that the probability of failure model can be linked with any post-flashover fire severity model in which the heat output can be defined in terms of gas temperature and emissivity.

10.5 The heat transfer submodel

The heat transfer model used in the research described in this chapter is fully described in Ref. [15]. The model has been named ADIDRAS, which is an acronym for Alternating Direction Implicit and Discrete Radiation Analysis (of heat transfer) in Structures. It uses a finite difference method of analysis for conductive heat transfer through solid materials. It also uses discrete radiative

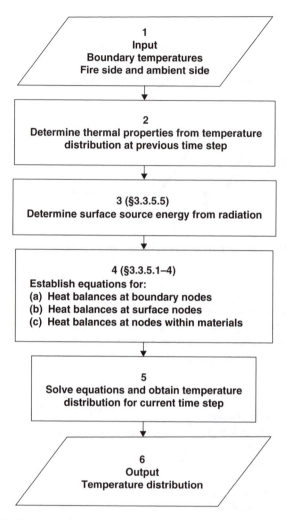

Figure 10.12 Heat transfer model – ADIDRAS.

heat transfer analysis for radiation through cavities. The model can be applied to a wide range of walls, including ordinary cavity walls, double-stud walls and staggered stud walls. Convective heat transfer at surfaces is modelled approximately with heat transfer coefficients. The thermal properties used in the heat transfer submodel are given in Table 10.1. A flowchart of the heat transfer model is shown in Figure 10.12.

10.6 The structural response submodel

The structural response submodel is fully described in Ref. [71]. It uses the same numerical grid as the heat transfer submodel. It determines the heat-affected stiffness of each element of material bounded by the grid lines. The mechanical properties used in determining stiffness are given in Table 10.1. If an element of wood is charred, its temperature is greater than 300°C, and hence its elastic modulus (Figure 10.8) and its stiffness are zero. The structural submodel determines the overall flexural stiffness of wall cross-sections, using composite section theory. Beam-column line members are taken to have the same overall flexural stiffness. This line member model is analysed with a second order stiffness analysis that enables the analysis of buckling. The structural response submodel well satisfies the computational requirements for the repetitive random simulation of Monte Carlo analysis – fast computation and numerical robustness. Robustness is required to ensure that any extreme values generated by the simulation do not cause numerical instability resulting in the potential waste of many hours or even days of computation.

A flowchart of the structural response model is shown in Figure 10.13.

10.7 Distribution of failure times

10.7.1 Walls subjected to standard fires and real loads with controlled selection of studs

Two sets of one hundred simulations for a wall exposed to a standard deterministic fire were undertaken. The coefficients of variation for loads, mechanical and thermal properties were as in Table 10.1 except for the coefficient of variation of the thermal properties of the wood studs, which was taken as 0.02 for the first set of simulations and 0.10 for the second set. The studs were selected so that the coefficient of variation of the compression strength, the tensile strength and the elastic modulus was 0.013. The resulting cumulative probability distributions are shown in Figure 10.14.

10.7.2 Walls subjected to standard fires and real loads with uncontrolled selection of studs

One hundred simulations for a wall exposed to a standard deterministic fire were undertaken. The coefficients of variation for loads, mechanical and thermal properties were as in Table 10.1. In all of the simulations the mode of failure was structural collapse. The resulting cumulative probability distribution is shown as the dashed line in Figure 10.15. The plot from Figure 10.14 (coefficient of variation of thermal properties 0.02) is repeated for comparison. The coefficient of variation of the failure time was found to be 0.12.

10.7.3 Walls subjected to a real fire and real loads with uncontrolled selection of studs

One hundred simulations for a wall exposed to a Kawagoe and Lie fire (Section 10.4.2) with a fire load Q of $10 \, \text{kg/m}^2$ were undertaken. The coefficients of variation for loads, mechanical and thermal properties were as in Table 10.1. It is to be noted that with real fires there is a non-zero probability that the

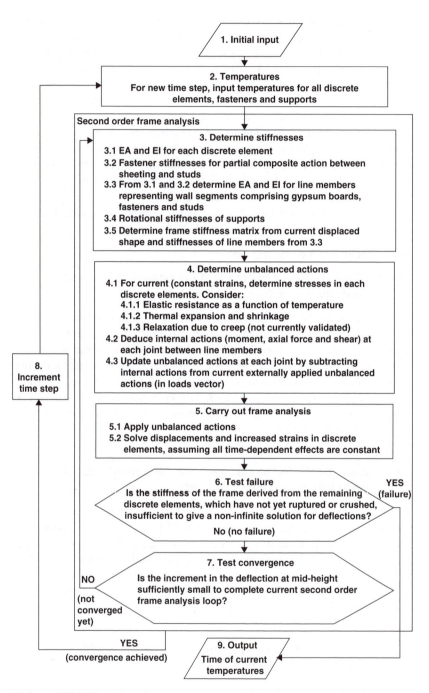

Figure 10.13 Flowchart giving overview of model for structural response.

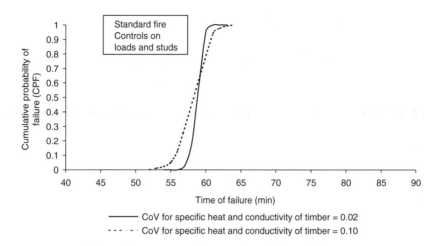

Figure 10.14 Cumulative probability of failure with time (standard fire and controls on loads and studs).

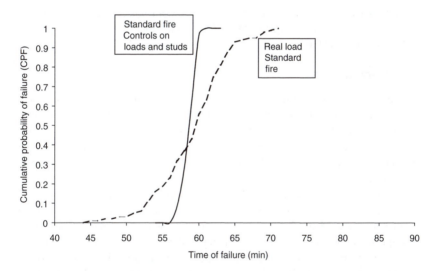

Figure 10.15 Comparison of the cumulative probabilities of failure with time for wood-framed wall with a realistic variation in mechanical properties and loads subjected to standard fire.

wall burn will not down. For the simulations carried out, the probability of failure was 0.61. In all of the simulations that resulted in failure the mode of failure was structural collapse. The resulting cumulative probability distribution is shown in Figure 10.16. The plots from Figures 10.14 and 10.15 are repeated for comparison. The coefficient of variation of the failure time, given that failure occurred, was found to be 0.12. This is similar to the coefficient of variation for standard fires. But this result was a coincidence.

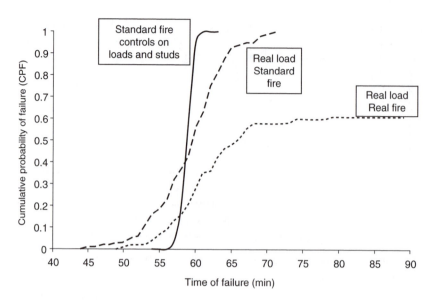

Figure 10.16 Comparison of the cumulative probabilities of failure with time for wood-framed wall with a realistic variation in mechanical properties and loads subjected to a real fire.

10.8 Conclusions

The research reported in this chapter was carried out to aid the application of new performance-based fire safety regulations to wood-framed wall construction. In the course of the research, coefficients of variation for thermal and mechanical properties of materials in wood-framed walls in fire were estimated. Unlike mechanical properties of wood, which are highly variable, the thermal properties appear to be remarkably consistent. For a given density, the coefficients of variation of the thermal properties of timber appear to be approximately 0.02 compared with the coefficients of variation for mechanical properties which range between 0.15 and 0.40. The reason for the consistency of thermal properties of timber appears to be the dependency of the properties on the constituents of wood, which are consistent: namely cellulose and lignin. The thermal properties appear to have little dependence on growth characteristics, such as knots and splits, which are responsible for the high variability of mechanical properties in wood. The practical implication of the low variation of thermal properties is that standard fire tests on loadbearing wood-framed walls can be made consistent and fair if studs are selected within a narrow range of elastic moduli.

Walls made from wood, which is randomly selected from graded supplies (Grade F8 [30,32]), leads to a significant coefficient of variation in the time to failure of 0.12. However, this variation is moderate because probability analysis has shown that walls, which are constructed in accordance with building regulations, will have fire resistance levels that are not significantly less than intended.

Advances in fire safety engineering should be focusing on, where possible, real fire scenarios. Insights into real fire scenarios have been given. Walls in real fire have some maximum probability of failure, less than 1.00, independent of time. That is, in reality and contrary to what may be assumed from standard fires, the occurrence of fire does not necessarily lead to walls collapsing. For the application demonstrated in this research, the variation of failure times, for those walls which failed, was similar to the variation failure times for the walls in standard fire. Further investigation is required to establish the variability of failure times of walls in real fire compared with failure times in standard fires in general.

Future research should be undertaken to increase the speed of the probability of failure model with the use of efficient numerical routines such as importance sampling functions to make probability analysis more convenient.

Due to the extensiveness of service penetrations and the suspected poor control of their fire resistance particularly during construction, surveys and research into the reduction of the times of failure due to service penetrations in walls and other construction details should also be undertaken.

11

A stochastic fire growth model

11.1 Conversion of a deterministic model to a stochastic model

In this chapter and Chapter 12 two models will be considered. One model describes fire growth and the other describes smoke spread. The stochastic version of the models is obtained by introducing randomness in the model.

The introduction of randomness in a deterministic model is discussed in Ref. [28]. Two methods are described. The first method, the more general one, consists in replacing the deterministic inputs to the model by random variables. The method implies that the major source of randomness is in the inputs, an assumption that is often realistic for fire models. The second method is to introduce random terms in the model equations, thus implying that the source of the randomness lies in the imperfection of the model. It is also possible to introduce randomness with both methods, as illustrated in Ref. [28].

The first method will be illustrated by the fire growth model. The second method will be illustrated by the smoke spread model in Chapter 12.

11.2 Description of the fire growth model

The fire growth model that will be analysed is called CESARE-Risk. It is based on the original article by Takeda and Yung [62], a recent report by Cooper and Yung [18], and the modification to the model by researchers at the Centre for Environmental Safety and Risk Engineering, at Victoria University of Technology (see He [34]). It has been used to generate temperature and smoke data for an integrated system model that incorporates many aspects of a building – fire situation, such as sprinkler and alarm response, smoke spread, human behaviour and egress, fire brigade response, structural failure, etc. [34]. It is used to predict the performance of building fire safety systems and to identify cost-effective fire safety system designs for buildings.

In this chapter an apartment fire growth model will be described. The model itself is deterministic. However, the model has been converted to a stochastic model by replacing the deterministic input variables by appropriate random ("stochastic") variables. These variables are described in Section 11.2.4.

The description of this model and its general assumptions are briefly as follows: the purpose of the apartment fire growth model is to simulate the ignition and growth of fires in an apartment unit in order to help assess the fire safety performance of apartment buildings. This assessment is done on the basis of the amount, temperature and concentration of the gases generated by the model fire, the speed and effectiveness of various fire protection systems, and the behaviour of building occupants.

The apartment fire growth model calculates the characteristics of compartment fires that have the greatest impact on occupant safety and building damage. These characteristics fall into two categories: the smoke and fire hazard category and the detection category. The former category includes the composition, temperature and flow rate of compartment effluent gases, and this information can be used to estimate the potential for smoke spread and fire damage outside the compartment of fire origin. The latter category includes the time of occurrence of specific fire-detection-related events, such as the time the person in the room of fire origin first notices the fire, the smoke detector activation time, the sprinkler activation time, the time of flashover and the time to fire burnout. These times can be used to determine the occupant response and evacuation time [34].

The model uses standard flexible polyurethane foams to represent the upholstered furniture and bedding typically found in an apartment. Flame spread and fire growth over other fuel materials can also be simulated. All physical parameters associated with foam may be changed to represent other fuels. However, the combustion chemistry is specifically formulated to describe the products of combustion produced by polyurethane under a range of enclosure conditions. The ventilation conditions simulated by the fire growth model include natural ventilation through door and window openings, which may be either open or closed, and forced ventilation from an air handling system, such as air conditioning or smoke extraction.

11.2.1 Assumptions

1. *One-zone model*
   - The ceiling, walls and floor of the compartment are fire separations.
   - The compartment is small (1–2 average size residential rooms).
   - The compartment gases are well mixed (at uniform temperature and pressure).
   - Flow through multiple compartment openings is weighted by area.
   - The compartment wall temperatures are uniform and equal.
2. *Heat transfer mechanisms*
   - Heat transfer by radiation, convection and conduction.
   - Heat transfer to fuel is assumed to occur mainly by radiation.
   - The floor is treated as an adiabatic boundary.
3. *Furniture arrangement*
   The furniture in the compartment is assumed to exist as a single mass in the centre of the room and to possess uniform properties. The size of the fuel mass reflects the amount of ignitable combustible in the room.
4. *Material properties*
   Material properties are assumed to remain constant at their ambient values except the gas density.
5. *Fire detection/suppression*
   The fire growth model does not calculate the effects of fire suppression.

Table 11.1 Stochastic input variables

Variable	Variable name	Symbol	Unit	Distribution	Interval
x_1	Length of room	L	cm	Uniform	(300, 1000)
x_2	Width of room	W_r	cm	Uniform	(300, 1000)
x_3	Height of room	H_r	cm	Uniform	(240, 300)
x_4	Window width factor	f_W		Uniform	(0.5, 1.0)
x_5	Window height factor	f_H		Uniform	(0.4, 1.0)
x_6	Fuel density	ρ_f	kg/m^2	Uniform	(20, 60)
x_7	Fuel area factor	f_A		Uniform	(0.3, 0.9)
x_8	Flame spread rate	R_f	m/s	Uniform	(0.1, 2.0)

$f_W = W_w/W_r$, where W_w is the window width in cm;
$f_H = H_w/H_r$, where H_w is the window height in cm;
$\rho_f = m_f \times 10^4/W_r L$, where m_f is the fuel mass in kg;
$f_A = \pi r_f^2/W_r L$, where r_f is the fuel radius in cm.

11.2.2 Scenarios

Four scenarios are considered: door open, window open; door open, window closed; door closed, window open; and door closed, window closed. They will be represented symbolically by DOWO, DOWC, DCWO and DCWC.

11.2.3 Output variables

The output variables of the model include: time to light smoke, time to medium smoke, time to heavy smoke, time to flare over, time to untenable conditions, maximum temperature reached and active time.

11.2.4 Stochastic input variables

The stochastic nature of the input is described in Table 11.1

The available data were obtained from a Monte Carlo simulation of size 10,000 with 2500 simulations for each of the four scenarios. The eight input parameters were assumed to be independent and their values were obtained by random sampling from the probability distributions specified in Table 11.1.

11.3 Response surface for maximum temperature

Regression analysis can be used to obtain a response surface (Section 3.14.4) representing the maximum temperature as a function of the input variables. We illustrate the procedure for the DOWO scenario. For full details of the derivation of the response surface see Ref. [30]. It can be shown that there are different modes of fire growth for room length L (i.e. x_1) less than 600 cm and room length greater than 600 cm. There are also different modes of fire growth for a flame spread rate R_f (i.e. x_8) greater than 0.455 and a flame spread rate less than 0.455. There were 1169 data sets satisfying the constraints $L > 600$ and $R_f > 0.455$ (excluding outliers). It also turns out that the room height x_3, and

Table 11.2 Values of quadratic regression coefficients

i	1	2	4	5	6	7
b_i	0.0910	0.0459	273.6	311.6	2.435	-154.1
a_i	-0.0001	0.000	-98.83	-125.1	-0.0183	33.80

the flame spread rate x_8, can be ignored in the regression calculations. So the set of indices used was just $I = 1, 2, 4, 5, 6, 7$.

To the 1169 data sets was fitted a quadratic regression formula of the form:

$$y = c + \sum_{i \in I} (b_i x_i + a_i x_i^2) + \epsilon. \tag{11.1}$$

The coefficient c was 743.23.

The b_i and a_i were as in Table 11.2.

Setting

$$y_t = c + \sum_{i \in I} (b_i x_i + a_i x_i^2) \tag{11.2}$$

it was found that the correlation between y and y_t was 0.969.

A second step in the fitting was to improve the fit of y_t to y by using a cubic regression formula of the form:

$$y = C_0 + C_1 y_t + C_2 y_t^2 + C_3 y_t^3 + \epsilon^*. \tag{11.3}$$

The coefficients turned out to be:
$C_0 = 2.547$,
$C_1 = -7.844$,
$C_2 = 9.915 \times 10^{-3}$,
$C_3 = -3.621 \times 10^{-6}$.

Letting

$$y_{\text{pred}} = C_0 + C_1 y_t + C_2 y_t^2 + C_3 y_t^3, \tag{11.4}$$

the correlation achieved between y and y_{pred} is now 0.972.

A scatter plot of y_{pred} against y is shown in Figure 11.1.

11.3.1 Calculation of the reliability index for engineering design

The definition of the reliability index β was given in Section 4.4.

The particular shape of the regression equation derived above makes the task of finding the design point and the reliability index extremely easy. For a fixed value of y_t the limit surface equation is

$$y_t = \sum_{i=1}^{n} a_i x_i^2 + \sum_{i=1}^{n} b_i x_i + c. \tag{11.5}$$

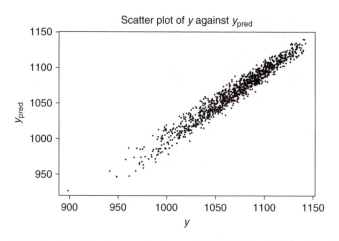

Figure 11.1 Scatter plot of y against y_{pred} for DOWO scenario, $L > 600\,\text{cm}$, $R_f > 0.455$.

Suppose the physical variables $\mathbf{X}$ are independent and normally distributed. For $i = 1, \ldots, n$ let X_i have mean μ_i and standard deviation σ_i, and let

$$u_i = \frac{x_i - \mu_i}{\sigma_i}.$$

Let the image of the limit surface in the $\mathbf{U}$ plane be

$$y_t = \sum_{i=1}^{n} A_i u_i^2 + \sum_{i=1}^{n} B_i u_i + C. \tag{11.6}$$

The design point is defined as the point on the limit surface nearest to the origin. In other words, we look for a vector $\mathbf{u}$ satisfying equation (11.6) that minimizes $\beta^2 = \sum_{i=1}^{n} u_i^2$. It can be easily found by using Lagrange's method of undetermined multipliers. Let

$$y^* = \lambda \sum_{i=1}^{n} u_i^2 + \sum_{i=1}^{n} A_i u_i^2 + \sum_{i=1}^{n} B_i u_i + C. \tag{11.7}$$

Then we must have

$$\frac{\partial y^*}{\partial u_i} = 2\lambda u_i + 2A_i u_i + B_i = 0 \tag{11.8}$$

from which we deduce that

$$u_i = -\frac{B_i}{2(\lambda + A_i)}. \tag{11.9}$$

Replacing in equation (11.6), we see that λ must satisfy the equation:

$$y_t = \sum_{i=1}^{n} \frac{A_i B_i^2}{4(\lambda + A_i)^2} - \sum_{i=1}^{n} \frac{B_i^2}{2(\lambda + A_i)} + C, \tag{11.10}$$

which is a polynomial equation in λ of order $2n$.

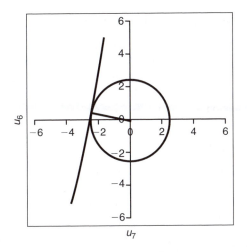

Figure 11.2 Illustration of the β index for the maximum temperature.

Due to the construction of the problem, this equation always has at least one real root. We choose the root that minimizes $\beta = \sqrt{\sum_{i=1}^{n} u_i^2}$. This minimum value of β is the required reliability index.

The corresponding probability of failure is given approximatively by $p_f = \Phi(-\beta)$.

11.3.2 A numerical example

For simplicity, we shall take just two input variables to be random: the fuel density ρ_f, denoted by x_6 and the fuel area factor f_A, denoted by x_7. The other input variables will be taken to be constant. This would usually be the case when dealing with a known building, since these other variables are just geometrical dimensions, apart from the flame spread rate R_f. However, we saw in Section 11.3 that the flame spread rate does not appear in the regression equation as long as it is larger than 0.455 m/s. Let $N(\mu, \sigma)$ denote a normal random variable with mean μ and standard deviation σ.

The assumed values are as follows: $L = 800$ cm, $W = 500$ cm, $H = 250$ cm, $f_W = 0.7$, $f_H = 0.5$, $\rho_f = N(30, 2)$, $f_A = N(0.6, 0.1)$ $R_f = 1.5$ m/s. The limiting state is taken to be $y_t = 1050°C$, corresponding to a value of $T_{max} = 1050.327°C$.

The regression equation now reduces in the **U** plane to

$$y_t = -0.0732u_6^2 + 0.3380u_7^2 + 2.674u_6 - 11.3563u_7 + 1018.848. \quad (11.11)$$

The value of λ that yields the smallest value of β is -2.6551 and the corresponding value of β is 2.4998. The approximate value of the probability of failure is $p_f = \Phi(-2.4998) = 0.00623$.

An illustration of the result is given in Figure 11.2.

11.3.3 Check by Monte Carlo simulation

The probability just given can be checked by Monte Carlo simulation: 200,000 simulations were carried out, of which 1222 fell in the failure region $y_t > 1050$. This gives an estimated probability of failure of 0.00611. The 95% confidence interval for the estimate is (0.00577, 0.00645), which contains the value obtained in the previous section.

11.3.4 Conclusion

Use of regression provides us with a simple response surface representation of the output of a computer fire model for an appropriate subrange of the input variables. This can in turn be used with minimal effort to obtain an accurate value for the reliability of a considered design, once the probability distribution of the input is decided. The reliability can be obtained by a first-order second-moment (FOSM) calculation or by Monte Carlo.

11.4 Calculation of the time to untenable conditions

Untenable conditions are assumed to occur either by the CO concentration reaching a level that results in incapacitation or if the occupant is exposed to an unbearable level of heat radiation. The time to untenable conditions is taken to be the time to reaching either of these conditions, whichever occurs first.

11.4.1 Calculation of COHb value

During evacuation under smoke conditions, occupants who are exposed to the smoke accumulate a carboxyhaemoglobin (COHb) dosage in the blood, through inhaling CO and CO_2. Occupant incapacitation will occur when the contents of COHb in the blood exceed a critical level estimated to be 20% of total haemoglobin. Fatality will occur when the critical level reaches 50%.

An equation (derived from experimental human exposures) for the prediction of COHb concentration is given by Stewart et al. [60]:

$$\%COHb = \int_0^t 3.317 \times 10^{-5} \times CO(t)^{1.036} \times RMV \times dt, \qquad (11.12)$$

where $CO(t)$ is the CO concentration in ppm as a function of time, RMV is the volume of air breathed (l/min) and t is the time of exposure (min).

The toxic gases considered in CESARE-Risk are CO and CO_2 only, because for most practical situations the composition of the fire atmosphere is such that the toxic effects of CO are the most important. The effect of CO_2 is calculated by the fire growth model. The CO_2 concentration is needed to determine a factor by which the COHb from CO is multiplied to take into account the increase of the breathing rate caused by CO_2. This factor is calculated using

$$V_{CO_2} = \exp(0.2468 \times CO_2\% + 1.9086)/6.8, \qquad (11.13)$$

where V_{CO_2} is the multiplication factor for CO_2 induced hyperventilation.

Thus, the total COHb with the effect of CO_2 is

$$\%\text{COHb} = \int_0^t 8.2925 \times 10^{-4} \times \{CO(t)\}^{1.036} \times dt \times V_{CO_2}, \qquad (11.14)$$

where 25 l/min is used for the rate of breathing which is the breathing rate for adults with light activity.

11.4.2 Fatality caused by heat

The second cause of incapacitation, exposure to heat radiation in a building compartment, will now be presented.

The most important sources of heat are radiative heat from the fire and convective heat from the hot gases. According to Babrauskas [7], the tenability limit for radiative heat flux Q_r is 2.5 kW/m^2 (0.25 W/cm^2). The radiative heat flux is closely related to the temperature. For simplicity, only temperature is used for determining the occupant fatality condition.

For exposure to convective heat, the concept of fractional lethal dose (FLD) can be used to predict whether a fatality will occur. FLD is defined to be

$$\text{FLD} = \frac{\text{Dose received at time } t}{\text{Dose to cause fatality}}. \qquad (11.15)$$

The FLD due to convective heat (FLD_T) is calculated using:

$$\text{FLD}_T = \int_0^t dt \Big/ \left(199.3 \times \exp\left(\frac{60.89 - T(t)}{4.18}\right) + 52 \times \exp\left(\frac{60.89 - T(t)}{29.83}\right) \right),$$

$$(11.16)$$

where $T(t)$ is the temperature of the hot gases as a function of time; t is the time at which the FLD is calculated.

A value of FLD_T greater or equal to 1 is assumed to be fatal to the occupants. This equation is derived from the data presented in Purser [51] using curve fitting techniques. Details of the above calculation in CESARE-Risk model can be found in Sanabria and Li [57].

It is important to notice that there are basic differences between maximum temperature and the time to untenable conditions from the point of view of reliability. Firstly, while the unsafe region for maximum temperature is the region where the maximum temperature exceeds some critical value, the unsafe region for time to untenable conditions is the region where the time to untenable conditions is shorter than the critical value. Secondly, the time to untenable conditions is inherently non-negative and moreover there is evidence that it tends to have a lognormal distribution (see Ref. [29]). It therefore makes sense to attempt to regress the logarithm of the time to untenable conditions on the input variables. This is carried out in Section 11.4.3. It is also interesting to note that the multiple correlation achieved between the time to untenable conditions

Table 11.3 Values of quadratic regression coefficients

i	1	2	3	7	8
b_i	0.0008228	0.0009558	0.002149	0.04475	−1.293
a_i	-2.8302×10^{-7}	-3.5024×10^{-7}	-2.3356×10^{-6}	0.01545	0.2913

and the transformed input variables was uniformly better than for the maximum temperature. Also, while some outliers were detected during the fitting of the maximum temperature to the input variables, no outliers were detected for the time to untenable conditions.

11.4.3 Logarithm of time to untenable conditions

In this section, attention is focused on the regression of the logarithm of the time to untenable conditions for the scenario DCWC. For full details of the derivation of the response surface see Ref. [32]. The regression equation derived is valid for $R_f > 0.5$. In this section, T will represent the time to untenable conditions.

There were 1988 data points satisfying the constraint $R_f > 0.5$. Moreover, it turned out that x_4, x_5, x_6 (window width factor, window height factor and fuel density, respectively) could be ignored in the regression calculations.

To the 1988 data points was fitted a regression formula of the form:

$$\log(T) = c + \sum_{i \in I} (b_i x_i + a_i x_i^2) + \epsilon, \qquad (11.17)$$

where $I = 1, 2, 3, 7, 8$.

The coefficient c was 5.240082.

The b_i and a_i were as in Table 11.3.

Setting

$$\log(T_t) = c + \sum_{i \in I} (b_i x_i + a_i x_i^2) \qquad (11.18)$$

$$T_t = \exp \left(c + \sum_{i \in I} (b_i x_i + a_i x_i^2) \right) \qquad (11.19)$$

it was found that the correlation between T_t and the original value T was 0.9984.

The second step in the fitting was to improve the fit of T_t by using a cubic regression formula of the form:

$$T = C_0 + C_1 T_t + C_2 T_t^2 + C_3 T_t^3 + \epsilon^*. \qquad (11.20)$$

The coefficients turned out to be $C_0 = -8.501$, $C_1 = 1.171$, $C_2 = -0.000939$, $C_3 = 1.4804e{-}006$.

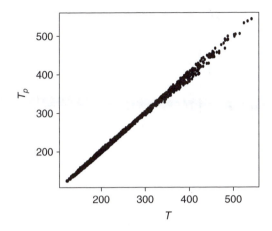

Figure 11.3 Scatter plot of T against T_t for DCWC scenario, $R_f > 0.5$.

Letting

$$T_p = C_0 + C_1 T_t + C_2 T_t^2 + C_3 T_t^3. \tag{11.21}$$

The correlation between T_p and the original value T is 0.9990. A scatter plot of T against T_p is shown in Figure 11.3.

11.4.4 Calculation of the reliability index for the DCWC scenario (logarithmic fit)

In the DCWC scenario, we just take two input variables to be random: fuel area factor W_r, which is denoted by x_2 and flame spread rate R_f, which is denoted by x_8. The other input variables are taken to be constant. This would be the case when dealing with a known building, since the other variables are just geometrical dimensions. Let $N(\mu, \sigma)$ denote a normal random variable with mean μ and standard deviation σ.

The assumed values are as follows: $L = 600$, $W_r = N(450, 20)$, $H_r = 250$, $f_W = 0.7, f_H = 0.6, f = 50, f_A = 0.6, R_f = N(0.7, 0.1)$.

The limiting state is taken to be $T_{p\min} = 230$ s. The corresponding value of $T_{t\min} = 230.9$, and the regression equation reduces in the **U** plane to

$$\log(230.9) = -0.0001401 u_2^2 + 0.01281 u_2 + 0.0029131 u_8^2$$
$$- 0.08847 u_8 + 5.653. \tag{11.22}$$

Using the methodology described in the previous section, the values of λ are:

λ: 0.0001, 0.0001, -0.0013, 0.0146,

and the corresponding values of β are

β_L: 73.5681, 73.5681, 27.3690, 2.5645.

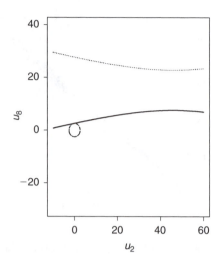

Figure 11.4 Illustration of the β index for the logarithm of the time to untenable conditions.

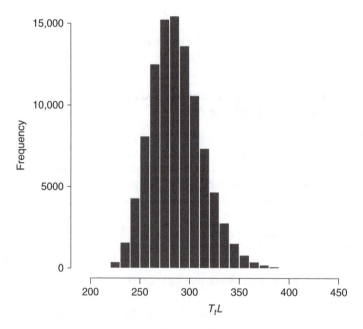

Figure 11.5 Histogram of the Monte Carlo simulation for the time to untenable conditions.

The smallest value of β is 2.5645 and this is the reliability index.

The approximate value of the probability of failure is $p_f = \Phi(-2.5645) = 0.00517$.

An illustration of the result is given in Figure 11.4.

11.4.5 Validation by Monte Carlo simulation

The probability just given by the reliability index can be validated by Monte Carlo simulation: out of 100,000 simulations, 529 observations fell in the failure region $T_p < 230\,\text{s}$ ($T_t < 230.9\,\text{s}$), giving an estimated probability of failure of 0.00529. The histogram is shown in Figure 11.5. The 95% confidence interval for the estimate is (0.00484, 0.00574), which contains the value (0.00517) obtained by using the reliability index in the previous section.

12

A stochastic model for smoke spread

12.1 Introduction

In this chapter, as discussed in the introduction of Chapter 11, the spread of smoke in a building is simulated by transforming a deterministic model into a stochastic model through adding random terms to the model equations. The underlying deterministic model is presented in Ref. [35]. The stochastic model is more fully described in Ref. [46].

The building considered for the simulation is a 10-storey building with a stairwell, corridors and compartments. The simulation is limited to spread in the stairwell, corridors on the floors and between levels.

The National Fire Protection Association [2] defines smoke as consisting of the airborne solid and liquid particulate and gases evolved when a material undergoes pyrolysis or combustion, together with the quantity of air that is entrained or otherwise mixed into the gases. The predominant hazards to humans exposed to products of combustion include heat, visual obscurity and narcosis. Narcosis results from the inhalation of asphyxiants and irritation of the upper respiratory tracts. Delays due to visual obscurity, panic and disorientation results in further inhalation of toxic gases.

12.2 Deterministic modelling

Since the conditions vary along a large enclosure, it is reasonable to divide it into a number of finite volumes and apply an assumption of uniform condition to each individual volume. Each individual volume can then be treated as a node in a network. The connections between nodes may represent openings or imaginary boundaries. This network approach is capable of predicting the temporal and spatial variations of the quantities of interest in large enclosures and yet retain the merit of computation speed.

The following modelling assumptions are made:

1. All quantities of interest are uniformly distributed within a volume.
2. The specific heat at constant pressure of the gas is independent of temperature.
3. The variations in the average molecular weight of the gas are insignificant.
4. The rate of pressure change is negligible.
5. The effect of soot on gas properties is negligible.
6. There is no depletion, generation or absorption of mass or species within the volume.
7. Friction loss is negligible.

The mass flow rate from one volume to the next is calculated from the consideration of mass conservation and thermal expansion.

The basic equations for the analysis are:

1. the identity,
2. the equation of state,
3. conservation of mass,
4. conservation of enthalpy,
5. conservation of species.

After some mathematical manipulations another set of equations can be derived that expresses the mass flow rate and the rate of variation of temperature and species concentration explicitly:

$$m_o = \frac{(m_i T_i - Q/C_p)}{T} \tag{12.1}$$

$$\frac{dT}{dt} = \frac{RT}{PV} \left[m_i(T_i - T) - Q/C_p \right] \tag{12.2}$$

$$\frac{dY}{dt} = \frac{RT}{PV} m_i(Y_i - Y) \tag{12.3}$$

where
m = mass flow rate (kg/m^3);
T = temperature (K);
Q = heat loss rate (W);
C_p = specific heat of constant pressure (J/kg K);
P = pressure (Pa);
R = gas constant = 8.32×103 (J/kg K);
V = volume (m^3);
Y = species mass fraction;
t = time (s).

Subscripts:
i = inflow;
o = outflow.

There are a number of additional refinements to the model. They were not modified in the conversion to a stochastic model. The reader is referred to Ref. [35] for details.

12.3 Validation of the model

A smoke spread experiment was carried out in 1993 at the National Fire Laboratory of the National Research Council of Canada (NRCC) by Hokugo, Yung and Hadjisophocleous [37]. Figure 12.1 is a simplified sketch of the NRCC smoke tower and the network representation for the modelling.

For the purpose of the present model, only the smoke spread in the stairshaft and corridors are of interest. Hence, only these two types of structure are presented in the figure. The floor levels are denoted by F1, F2, . . . , F10. Each

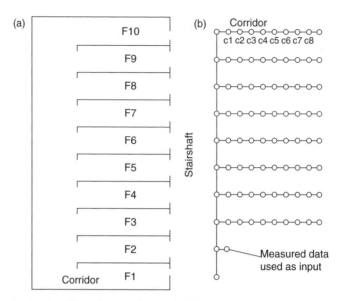

Figure 12.1 Simplified sketch of the NRCC smoke tower (a) and a network representation (b).

corridor was divided into nine finite volumes which are denoted by $c1, c2, \ldots,$ $c9$. The height of the first and second floor was 3.6 m and the height of all other floors was 2.6 m. The width and depth of the stairshaft were 5.05 m and 2.48 m, respectively. The length and width of the corridors were 14 m and 1.5 m, respectively. The walls of these structures were constructed of 200-mm-thick monolithic reinforced concrete. The free area of the stairdoor on each level was 0.16 m^2 and the effective area of the opening from corridor to outside was estimated to be 0.07 m^2. Fire was set on the second floor with two propane burners. Fresh air for the burners was piped in from the outside. Temperatures and carbon dioxide concentrations were measured along the stairshaft at the elevations of every floor and in each corridor near the door to the shaft. The measured smoke exhaust rate across the apartment door, temperature and carbon dioxide concentration on the floor of fire origin were used as input to the programme.

There were discrepancies between the modelled and the measured temperatures and species concentrations. They can be attributed to:

1. the inaccuracies in the measured data;
2. the finite volume approximation;
3. the inherent randomness of the smoke spread phenomenon, due among others, to the turbulent behaviour of the hot gases.

12.4 The stochastic model

Following the method described in Ref. [28], the deterministic model described by equations (12.1)–(12.3) is transformed into a stochastic model by adding random terms.

The form of the random terms is dictated by the methodology of *stochastic*, or *random differential equations*, as explained in Ref. [28]. The random terms are derived from the so-called *Wiener process*. The Wiener process is a random function of time, usually denoted by $W(t)$, that has the property that all increments $\delta W = W(t + \delta t) - W(t)$ are normally distributed with zero mean and variance δt. Moreover, all non-overlapping increments are independent of each other. It can be shown (see e.g. Soong [59]) that these properties of the Wiener process are consistent. The Wiener process is introduced into the equations as a *differential* dW.

Here three independent Wiener processes are introduced: $W_1(t)$, $W_2(t)$ and $W_3(t)$. Three *forcing functions* f_1, f_2 and f_3 are also introduced as scale parameters for the Wiener differentials. Since f_1 is introduced in equation (12.1) which does not contain the variable P, it is taken as a function of T only. The other two forcing functions are functions of T and P.

The purpose of introducing the forcing functions is to model the intrinsic variability of the fire phenomenon.

In addition, equations (12.1)–(12.2) are written in differential form, so that the stochastic model is described by the following three stochastic equations:

$$\left[\frac{m_o - \left(m_i T_i - \frac{Q}{C_p} \right)}{T} \right] dt = f_1(T) \, dW_1. \tag{12.4}$$

$$dT - \frac{RT}{PV} \left[m_i(T_i - T) - \frac{Q}{C_p} \right] dt = f_2(T, P) \, dW_2. \tag{12.5}$$

$$dY - \frac{RT}{PV} m_i(Y_i - Y) \, dt = f_3(T, P) \, dW_3. \tag{12.6}$$

The three forcing functions were chosen so as to bridge the difference between the computer model and the observations. They were taken as:

$$f_1(T) = \phi T / 10^5, \tag{12.7}$$

$$f_2(T, P) = 50\phi TP / 10^7, \tag{12.8}$$

$$f_3(T, P) = 12\phi TP / 10^7, \tag{12.9}$$

where T and P are, respectively, temperature and pressure. The parameter ϕ was chosen to fulfil the purpose of introducing stochasticity, which is, as mentioned above, to reconcile the computer model with the observations. It was found by trial and error that a suitable value was 200.

12.5 Results of the simulations

Equations (12.4)–(12.6) were suitably discretized and then used to carry out a Monte Carlo simulation of the smoke spread. The sample size was 1000. As previously mentioned, the purpose of the simulation was to reconcile the results of

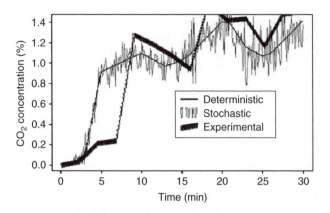

Figure 12.2 Comparison of the deterministic and stochastic models for gas (CO_2) temperature on level 2 stairway with Hokugo's results.

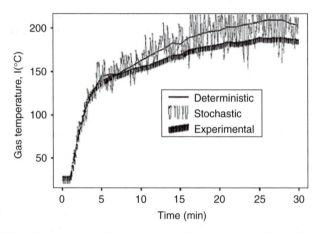

Figure 12.3 Comparison of the deterministic and stochastic models for CO_2 concentration on level 2 stairway with Hokugo's results.

the deterministic model with the experimental observations. The reconciliation is illustrated in Figures 12.2 and 12.3.

Both figures refer to the level 2 stairway and show the time-dependent deterministic model output, the experimental observations and one simulation of the stochastic output. Figure 12.3 refers to the CO_2 concentration while Figure 12.3 refers to the gas temperature. It can be seen that the stochastic modification of the deterministic model bridges the differences between the deterministic computer model and the observations.

The success of the reconciliation can be further tested statistically as follows. Consider just the 30th minute from the start of the fire at the stairway of the second floor. The Monte Carlo simulation generates a sample of size 1000 of the gas temperature. A histogram of the sample is given in Figure 12.4.

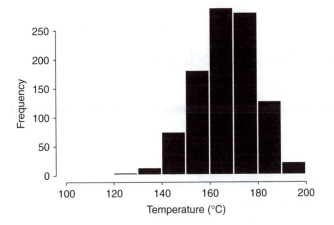

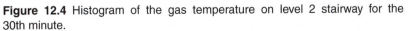
Temperature (°C)

Figure 12.4 Histogram of the gas temperature on level 2 stairway for the 30th minute.

The mean is 167.42°C and the standard deviation is 12.78°C. The observed value was 173.45°, which lies at less than one standard deviation from the simulation mean. The proportion of sample values larger than the observed value is given by

$$\Phi\left(\frac{173.45 - 167.22}{12.78}\right) = 0.313. \qquad (12.10)$$

Similar tests can be carried out for different times, different locations and different outputs, e.g. CO_2 concentration.

13

A stochastic model for human behaviour

13.1 Overview of the model

The model described in this chapter was developed within the framework of the Fire Code Reform Centre Project 4 [14]. It is a submodel of the CESARE-Risk Model. Its operation starts at fire initiation. The duration of operation is a user input variable. The submodel calculates the number of persons remaining in different locations in a residential building at particular times during a fire incident and calculates the incapacitation of occupants based on a fractional incapacitating dose (FID) formulation and heat radiation. The ultimate aim of the submodel is to calculate the expected number of deaths (END) for people who remain in the building.

The model has input from other CESARE-Risk models, in particular the system model and the smoke and fire growth models.

Occupant response in a fire emergency can be depicted as involving three stages (recognition, action or coping behaviour, and evacuation). These stages are a convenient summary of many actions, the most diverse being in the coping stage, and they are not universally applicable. In order to cope with the operational demands of a time-based risk model, the time when evacuation starts is used to separate the first two stages (initial recognition and coping behaviour) from the third stage (evacuation movement), leading to the distinction between the response model and the evacuation model.

Within the two periods, there are fixed units of time. The time step from the system model is the default time unit for the human behaviour model. The model runs on an increasing time orientation, defined as the time length from fire initiation counted in the time unit.

Variation in the time of cue occurrence, and in the probability of recognition and predicted speed of response by different occupants means that the two submodels are interdependent. The evacuation model interacts directly with the response model when it is used to determine the time warnings are issued.

The response model deals with behaviour of occupant groups from the time of fire initiation up to the time when evacuation begins with the occupant leaving an apartment. Thus it merges the recognition and coping stages. It links the direct and indirect indications of the presence of a fire in a building (cues) which are available to occupants to the response of the occupants. It takes into account the probability and time of occurrence of different cues, applies probabilities of response to different cues for different occupant groups, applies times for delays before occupants start to evacuate, and determines the number of people in different apartments who leave the apartments at particular units of time. Probabilities and times are supported by a Response in Fires database.

The evacuation model deals with the movement of people outside their apartments until they exit the building. It calculates the number of persons in different locations at nominated times during a fire incident and the time occupants have spent in each location. This information is used to calculate the END for people who remain in the building using a FID approach based on the link between the time occupants have spent in each location and the amount of heat or CO and CO_2 present in each location.

In the operation of the human behaviour model, the building is divided into three major locations which are further divided:

1. LFO, level of fire origin, which consists of:
   (a) apartment of fire origin: divided into RFO, room of fire origin, and RNFO, room not of fire origin;
   (b) apartment(s) of non-fire origin;
   (c) corridor;
   (d) stairways.
2. Levels above the LFO, which consist of:
   (a) apartment of non-fire origin apartments on each floor above LFO;
   (b) Corridor on each floor above LFO;
   (c) stairs on each floor above LFO.
3. Levels below the LFO, which consist of:
   (a) apartment of non-fire origin apartments below LFO;
   (b) corridor on each floor below LFO;
   (c) stairs on each floor below LFO.

When the model runs, the following conditions or principles operate:

1. All occupants are assumed to start from their apartments (or rooms in Class 3 buildings). Apartment of fire origin occupants begin either in the room of fire origin (RFO) or elsewhere in the apartment (RNFO) and cannot be in both. Apartment of non-fire origin occupants are distinguished according to whether they are in apartments on the level of the fire, above it or below it.
2. Apartment of non-fire origin occupants belonging to the same occupant group are aggregated and treated as one on any one level. Response is determined as an expected value for an occupant group. For the calculation of recognition probabilities and of probabilities of movement from an apartment (i.e. starting evacuation), and of the times at which such movement occurs in response to any one cue, occupants belonging to the same occupant group on any one level retain their group and apartment identity.
3. The response model establishes the cue occurrence sequence for apartment of non-fire origin apartments on each level and for RFO or RNFO occupants in the apartment of fire origin. For any apartment cell a_{ij} in a building, there are n cues which occur.

4. Cues occur at one time point. Occupants cannot respond to a cue which occurs while they are actively responding to an earlier cue.
5. Three times with different probabilities are used to characterize the distribution of the time taken to start evacuation.
6. The focus of the human behaviour model is on whether occupants move from one location to another and on the time delay involved, so it is concerned only with predicting evacuation and with response during evacuation. Occupants cannot affect fire growth or the movement of smoke through their actions.

The key steps in the model are summarized in Figure 13.1.

13.2 Historical note

The systematic study of human behaviour during fires began in the 1970s. Studies of human behaviour across a number of fires and occupancy types (Wood [70], Bryan [8], Canter *et al.* [12], Haber [25] and investigations into particular fire incidents (e.g. Bryan [9], Abe [4], Bryan [10], Donald and Canter [21], Scanlon [58], Swartz [61]) have demonstrated that patterns of behaviour can be identified in fire emergencies. It is this predictability that forms the basis for the modelling of human behaviour.

13.3 Occupancies and occupant distribution

The model was developed to cover various types of residential buildings. However, in this chapter, only the details of the application of the model to apartment buildings will be given, in order to illustrate the model methodology.

13.3.1 Defining the residential building

The residential building as defined by the system model is composed of various building cells: apartments, corridors, stairways and an elevator. The specification for a building includes:

1. Number of levels
2. Number of apartments on each level
3. Number of stairways on each level (less than four)
4. Geometrical size ($W \times L$) of each apartment, corridor and stairway
5. Door width (W_d) between two cells (if applicable)
6. Travel distance between two cells (D)
7. Elevator specification (cell available but function not described)
8. Assumed LFO.

13.3.2 Occupant profiles

Individually each residential apartment building is likely to accommodate a unique cross-section of occupants. However when the occupants of all residential apartment buildings within the country are considered as a whole, a typical

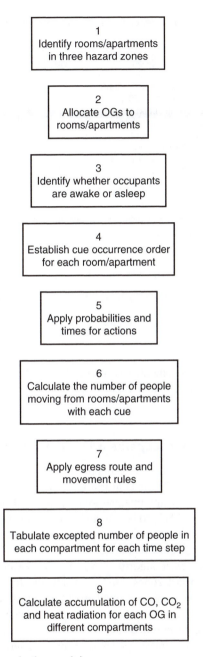

Figure 13.1 Key steps in the model.

occupant type and group distribution emerge. This allows a typical apartment population profile for a generic apartment building to be established.

Demographic data mostly obtained from the Australian Bureau of Statistics were used to establish occupant groups representative of occupants of Australian

Table 13.1 Occupant group description and percentage representation (day and night scenarios.)

Occupant group	Brief description	Number in room	% day	% night
1	Lone person <71 years	1	35	33
2	Lone person >70 years	1	8	6
3	2 Parents, child	3	20	20
4	2 Persons, no children	2	31	31
5	Lone person incapacitated by drugs/alcohol	1	1	5
6	2 Persons >70 years one with handicap	2	5	5

Table 13.2 Hazard zones in a residential apartment building

Hazard zone	Code	Location
Hazard zone 1	I_{LFO} and $(I_{LFO} + 1)$	LFO and level above (LFO only if top floor)
Hazard zone 2	$(I_{LFO} + 2)$, $(I_{LFO} + 3)$, etc.	Levels 2 or more floors above LFO
Hazard zone 3	$(I_{LFO} - 1)$, $(I_{LFO} - 2)$, etc.	Levels below the LFO

apartment buildings in general. Analysis of the available census data resulted in the occupant groups listed in Table 13.1.

13.3.3 Allocation of occupants within the apartment building

For the purpose of allocating occupant groups, three hazard zones are identified as shown in Table 13.2. These conform to the fire growth and smoke spread models. It is assumed that apartments/rooms on the LFO and on the level immediately above are more likely to be affected by the immediate hazards of the fire and smoke. Hazard reduces on floors two or more levels above the LFO, and is insignificant on floors below that level.

The model uses a basic algorithm in order to allocate occupant groups to apartments in all hazard zones, first adjusting the hazard zone boundaries to make sure there are sufficient number of apartments in each hazard zone to run the algorithm. The flowchart in Figure 13.2 demonstrates the hazard zone boundary adjustment procedure.

The number of apartments in each hazard zone is determined and the number of apartments for each group is allocated according to the occupant group relative frequency as given by Table 13.1. The numbers obtained are then rounded up to an integer value, but keeping the sum still equal to the total number of apartments in the hazard zone.

Finally, occupant groups are randomly allocated into each apartment in each hazard zone.

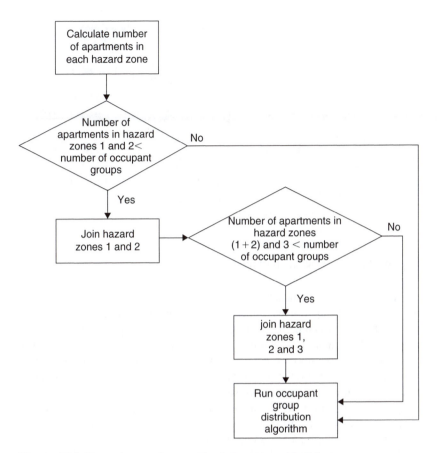

Figure 13.2 Hazard zones in a residential apartment building.

13.3.4 State of alertness

Occupant state of alertness (awake or asleep) is considered by having separate runs reflecting day and night conditions.

13.3.5 Occupancy rate

In programme operation, the occupancy rate for a building can be adjusted from 100% to allow for a lower occupancy rate during the day.

13.4 Response model

13.4.1 Introduction

The human behaviour model distinguishes occupant groups and flags changes to their status and location throughout the programme.

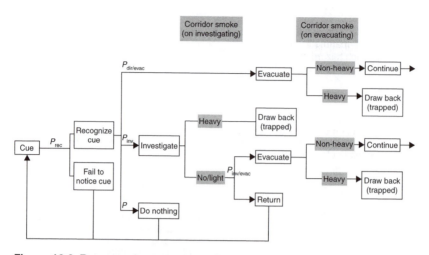

Figure 13.3 Response in apartments of non-fire origin.

The response model deals with behaviour up to the start of the evacuation process when occupants leave their rooms or apartments intending to leave the building. This is a period of much diversity and complexity as occupants recognize threat and make initial attempts to cope with a fire emergency. The response model focuses on predicting whether and when occupants move from their initial location into the corridor in response to defined cues. It uses recognition probabilities and action probabilities to calculate the expected number of occupants moving from rooms or apartments at each time step. Times for response distinguish between those who start evacuation without investigating and those who investigate prior to evacuation.

Probabilities for cue recognition are estimated using evidence from varied sources, with some support from a database on response in real fires. The database provides probabilities for action as well as for the action delay times.

The response model takes into account whether occupants are initially awake or asleep and assumes that a sleeping occupant who has woken to one cue will be awake on receipt of the next cue.

Occupants in an occupant group on any particular level are treated as a unit when probabilities and times are applied, with the exception of the apartment of fire origin.

13.4.2 Occupant response

Figure 13.3 indicates how the response of occupants in apartments of non-fire origin is conceptualized and Figure 13.4, the response of occupants in rooms and apartments of fire origin.

Response in the apartment of fire origin is simpler than in the apartments of non-fire origin as the occupants who recognize any of the designated cues do not choose to wait for further information. If they are in the RFO they will start evacuation. If they are in the RNFO they will start evacuation after locating the RFO.

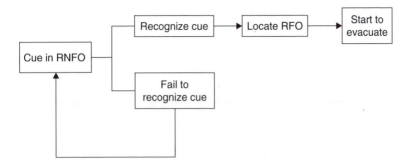

Figure 13.4 Response in rooms and apartments of fire origin.

Table 13.3 Probabilities of cue recognition when awake (OG: occupant group)

Cue	OG1, OG3 and OG4	OG2 and OG6	OG5
Light smoke	1.00	1.00	0.00
Local alarm	0.99	0.95	0.00
Corridor alarm	0.78	0.72	0.00
EWIS	0.90	0.90	0.00
Warning	1.00	1.00	0.00
Staff instruction	1.00	1.00	0.00

Table 13.4 Probabilities of cue recognition when asleep (OG: occupant group)

Cue	OG1 and OG4	OG3	OG2 and OG6	OG5
Light smoke	0.10	0.07	0.10	0.00
Local alarm	0.98	0.66	0.91	0.00
Corridor alarm	0.73	0.49	0.67	0.00
EWIS	0.80	0.60	0.80	0.00
Warning	1.00	1.00	1.00	0.00
Staff instruction	1.00	1.00	1.00	0.00

The difference between the RFO occupant response and the RNFO occupant response and the time for their response depends on the time of arrival of the cue.

13.4.3 Probabilities for recognition and action prior to starting evacuation

Probabilities for recognition and for action are calculated on the basis that cues are independent. For those occupants who choose to remain in the room or apartment after recognizing a cue, any subsequent cue is treated as though it is the first cue. Consequently, the probability for starting evacuation by occupants who have recognized an earlier cue is regarded as conservative.

Probabilities for cue recognition are summarized in Table 13.3 for occupants when awake and Table 13.4 for occupants when asleep. Where there is more

Table 13.5 Probabilities for initial action given cue recognition (apartment of fire origin)

Location	Probability
ROF: start evacuation	1.00
RNFO: locate fire then start evacuation	1.00

Table 13.6 Probabilities for initial action given cue recognition (apartment of non-fire origin)

Cue	Investigate	Start evacuation	Do nothing (wait)
Light smoke	0.50	0.00	0.50
Local alarm	0.80	0.10	0.10
Corridor alarm	0.28	0.12	0.60
EWIS	0.05	0.90	0.05
Warning	0.00	1.00	0.00
Staff instruction	0.00	1.00	0.00

than one person in a room it is assumed that other occupants will be woken if one occupant recognizes a cue.

Probabilities for initial action given cue recognition are given in Table 13.5 for the apartment of fire origin and Table 13.6 for apartments of non-fire origin. The acronym EWIS stands for Early Warning Intercommunication System.

13.4.4 Time of cue occurrence

The time of occurrence of cues for the response model is inputted from a fire growth and smoke spread models (Table 13.7).

13.4.5 Response times

1. The time to recognize a cue (occupants asleep only) is taken to be 30 s.
2. The time to investigate (RNFO only) is taken to be 30 s for occupants awake and 45 s for occupants asleep.
3. The times until the start of direct evacuation are given as a three-point discrete probability distribution (Table 13.8: occupants awake, Table 13.9: occupants asleep).
4. The times until completion of investigation and start of direct evacuation (apartment of non-fire origin only) are given as a three-point discrete probability distribution (Table 13.10: occupants awake, Table 13.11: occupants asleep).

Table 13.7 Cues available to apartment of non-fire origin and apartment of fire origin occupants

Initial location	Cues in room/ apartment	Smoke condition in corridor on investigating	Smoke condition in corridor on starting evacuation
Apartment of non-fire origin	Local alarm (apartment) Building alarm (corridor) Warning Light smoke Staff warning	No smoke Light smoke Heavy smoke	Heavy smoke Any other level of smoke or no smoke
Apartment of fire origin, RFO	Local alarm (apartment) Building alarm (corridor) Light smoke Staff warning		
Apartment of non-fire origin, RNFO	Local alarm (apartment) Building alarm (corridor) Light smoke Breaking glass Staff warning		

Table 13.8 Distribution of time to start evacuation (occupants awake)

Time (s)	Probability
12	0.22
62	0.59
121	0.19

Table 13.9 Distribution of time to start evacuation (occupants asleep)

Time (s)	Probability
46	0.46
267	0.39
963	0.15

Table 13.10 Distribution of time to investigate and start evacuation (occupants awake, apartment of non-fire origin only)

Time (s)	Probability
59	0.47
165	0.19
316	0.33

Table 13.11 Distribution of time to investigate and start evacuation (occupants asleep, apartment of non-fire origin only)

Time (s)	Probability
100	0.47
369	0.39
1266	0.14

Table 13.12 Inputs to the evacuation model

Input variables	Source
Building characteristics	
• Network definition of the building (nodes and arcs)	System model
• Building geometry (dimensions)	System model
• Egress routes	System model
Occupant characteristics	
• Occupant numbers at each node (initial and as updated by the response model)	Response model
• Baseline travel speed	Literature

13.5 Evacuation model

13.5.1 Introduction

The evacuation model of the human behaviour model deals with the behaviour after occupants leave their apartments. It predicts occupant locations in the building at any time after leaving their apartments and the condition of occupants. The time when occupants leave their apartments is obtained from the response model which is described in Section 13.4. Table 13.12 lists the inputs to the evacuation model.

13.5.2 Description of the model

The building is described as a building network which is composed of nodes and arcs. A node represents a physical building node such as an apartment, a corridor, a stairway or an exit. An arc represents a directed link between two nodes. The evacuation route for any occupant is determined by linking all arcs from the apartment cell to the building exit.

The model is based on the assumption that occupants are familiar with the layout of a building, a reasonable assumption for residential occupants. It assumes that occupants choose the shortest travel route leading to a building exit. The shortest route criterion can be replaced to allow a preference for a particular stair, to accommodate a familiarity criterion for example. The stairway preferential coefficient must be assigned at the beginning of the run of the model otherwise a default coefficient of $1/n$ (n is the number of stairways) is applied.

The total evacuation time for an occupant subgroup is the sum of the travel time through all arcs from the apartment to the exit. The programme examines all nodes at each time step and at each iteration moves occupants from the current start node to the current end node based on their travelling time. Time spent in travelling from one node to another has two components: traversal time and queuing time. Traversal time depends on the density of people in each node, type of node and occupant group. The evacuation model also deals with congestion occurring between and within any two building nodes.

The evacuation model takes into account the effects of the fire and its products on occupant response and the speed of that response. Alternative evacuation routes are modelled for circumstances when occupants confront smoke. Wood (70) and Bryan (8) among others report that many occupants are prepared to move through fairly dense smoke once they have decided to evacuate. The model assumes that when occupants encounter heavy smoke they try to avoid it by changing their direction if possible. Only heavy smoke is considered.

The model permits the use of alternative routes when there is heavy smoke in one or both stairways on any level. People do not continue on their evacuation route if they confront heavy smoke (with a visibility of less than 1 m) on reaching the stairs.

The exception for changing direction is in the case of an occupant who is leaving the apartment of fire origin. Where a corridor has heavy smoke, any occupant entering the corridor from the apartment of fire origin must continue to enter it, since the situation in the apartment of fire origin is always considered worse than in the corridor. The occupant who is in the corridor is deemed to be trapped by smoke. In contrast, apartment of non-fire origin occupants turn back if they confront heavy smoke as they begin evacuating. Occupants deemed trapped by smoke do not continue to move.

Balconies are nominated as places of refuge for apartment of non-fire origin occupants who recognize cues but who do not evacuate or who have had to return to their apartment. Once on a balcony, an occupant is assumed to be able to avoid smoke and heat. This outcome depends on building design rather than occupant response and may be overgenerous. In any case, apartment of fire origin occupants evacuate if they have recognized a cue: the model does not consider situations where a person might be unable to evacuate because of the actual location of the fire within the apartment of fire origin. In addition, occupants in apartment of non-fire origin apartments located on Levels 1 (ground) and 2 are assumed to reach safety by using windows as alternative exits.

The number of occupants at each location in a building at each time step is obtained by checking each node in a building network continuously for each time step. Once occupants move from their apartment they may be in the corridor, on stairs or have evacuated or, under the conditions above, have returned to their apartment.

The model thus describes occupants according to whether they have completed evacuation, whether they are moving through or incapacitated in corridors or stairways, or whether they are in apartments. People in apartments may or may not have recognized cues. For those who have, the model distinguishes people who did not attempt evacuation and people who were forced back into the apartment.

13.5.3 Incapacitation and fractional incapacitating doses

The model calculates the amount of CO and CO_2 absorbed by occupants from the time the fire starts until they evacuate, reach a balcony or are declared a fatality. It calculates a COHb value (based on Stewart *et al.* [60]), and applies critical levels derived from Purser (51) to determine incapacitation and death. The level for incapacitation is 20%. A level of 30% was under consideration but the conservative value was adopted for a number of reasons. They are that there is no allowance in the present model for differential effect according to the occupant group, or for reduced travel speed due to impaired vision or breathing difficulties in smoke, or for increased uptake of gases as a result of activity. The level for a fatality is taken to be 50%. Exposure to heat radiation results in death.

13.5.4 Categorization of occupants in the evacuation model

Occupants are regarded as either mobile or non-mobile. A mobile occupant may become non-mobile.

Non-mobile occupants are further divided into four categories which are mutually exclusive:

1. Trapped: those mobile occupants who do not evacuate but stay in their rooms because of smoke-logged corridors.
2. Disabled: who include the "incapacitated through drugs or alcohol" and the "dependent".
3. Incapacitated: those who are affected by gases but have not absorbed sufficient CO and CO_2 to reach a fatality level.
4. Dead: occupants who have been affected by radiant heat or have absorbed sufficient CO and CO_2 to reach a fatality level.

13.5.5 Output from the evacuation model

The output from the evacuation model is:

- Location of mobile and non-mobile people at each time step.
- Numbers of non-mobile people in four categories – trapped, disabled, incapacitated or dead – at each time step.
- Total evacuation time of the building.

14

Performance assessment of fire safety systems

14.1 Introduction

One of the problems in attempting rational design of fire safety systems for buildings is the development of an understanding of the relative effectiveness of fire safety systems that act in different ways. For example, comparisons of systems based on automatic sprinklers with those based on fire resistance ratings for structural elements and barriers are often frustrated, following a comprehensive comparison of the systems, by the question: "But what happens if the sprinklers fail?".

An assumption behind this question is that while sprinklers may sometimes fail, other systems do not. It appears that this attitude stems from the historical development of fire safety systems in building regulations and the like. These systems appear to have initially been largely based on two principles:

1. Separation
2. Use of non-combustible materials.

Separation was achieved by requiring certain distances between buildings or by requiring solid masonry walls of certain thicknesses and details if the buildings had to be close together.

Sprinklers did not exist at that time and thus could not be specified.

However, in recent times sprinklers and various other subsystems and components have been specified in building regulations and the like, usually in addition to the existing requirements.

To evaluate the effectiveness of each system, subsystem or component we can look at the changes in outcomes when the systems are present. For some fire safety systems, subsystems and components data that enables the effectiveness in relation to certain objectives to be evaluated can be found in fire statistics [3]. An alternative is to estimate the reliability (and to some extent effectiveness) of systems, subsystems and components not explicitly mentioned in the available fire statistics by the use of probabilistic risk assessment. The presentation in this chapter is based on the article by Thomas [64].

14.2 Using statistics to assess efficiency of safety systems

In the NFIRS database (see Section 8.2) there are fields identifying the presence or absence of sprinklers, detectors (whether smoke or heat detectors is not specified) and protected construction. There is however no indication of the presence, type or condition of alarm system, exit signage, portable extinguishers, fire safety plan (including evacuation plan) or management, smoke management system, etc. There are no details of building, room, corridor or stair dimensions or the ventilation in the room of fire origin or elsewhere, and only limited details of the fuel available to the fire.

Nevertheless, it is possible to obtain some indications of the value or effectiveness of sprinklers, detectors and protected construction. The effectiveness of systems, subsystems and components can be thought of as a combination of their *reliability* (the probability of operating as intended when required) and their *efficacy* (how well they achieve specific objectives given that they operate). Reliability is independent of objectives, efficacy (and therefore effectiveness) is not. For example, a sprinkler system in a building will have a certain reliability which will be the same no matter what objective is under consideration. However, the sprinkler system may have different levels of efficacy (and therefore effectiveness) when thought of as being intended to reduce property damage compared to its being intended to reduce civilian casualties or being intended to reduce fire fighter casualties.

There are also differences in the ability to estimate effectiveness depending on the objectives because of the frequency of occurrence of certain outcomes. The effectiveness in relation to the objective *reducing the spread of fire* is easier to estimate than the effectiveness in relation to an objective such as *reducing civilian injuries* which is in turn easier to estimate than the effectiveness in relation to *reducing civilian fatalities*. The reason for these differences is the frequency of occurrence of relevant outcomes. The extent of flame damage can be (and generally is) recorded for every reported fire, whereas civilian injuries occur in between 20 and 70 fires per 1000 reported fires for both residential and non-residential occupancies, and civilian fatalities occur in less than 20 fires per 1000 reported fires in residential occupancies, and generally in less than 1 fire per 1000 reported fires in non-residential occupancies.

Thus data is required on much larger numbers of fires for the estimation of effectiveness in relation to civilian fatalities than for the estimation of effectiveness in relation to civilian injuries and for both in comparison to the number required for the estimation of effectiveness in relation to extent of fire spread.

The data in the following tables are for years 1983–1995, excluding 1986, for apartments and 1983–1991 for offices.

Four cases are considered:

1. No fire safety system present
2. Protected construction
3. Detectors
4. Sprinklers.

For further comparisons including the effect of having more than one type of safety system see Ref. [64].

Table 14.1 Percentage of fires with extent of fire damage (apartment buildings)

Extent of flame damage	Number of fire safety system present	Protected construction	Detectors	Sprinklers
Single storey				
>Area	41.5	30.6	23.3	0.0
>Room	25.4	17.4	11.1	0.0
>Building	2.7	1.8	0.3	0.0
Number of fires	3301	2461	2900	9
2–4 Storeys				
>Area	38.4	34.7	27.0	19.8
>Room	25.6	21.9	15.8	6.6
>Storey	18.0	13.9	9.7	2.5
>Building	2.3	1.8	1.0	0.8
Number of fires	17,464	19,428	31,290	121
5–12 Storeys				
>Area	34.6	25.7	18.4	8.8
>Room	18.6	11.8	10.0	0.0
>Storey	12.3	3.5	5.9	0.0
>Building	0.9	0.4	0.8	0.0
Number of fires	587	1842	1597	34

14.3 Statistics of extent of fire damage

Table 14.1 gives the proportion of fires in apartment buildings with different number of storeys that result in a different extent of fire damage.

Examination of Table 14.1 reveals that protected construction reduces fire damage, that detectors are more effective than protected construction and that sprinklers are the most effective means to reduce fire damage. The data for office buildings (not reproduced here) reveal a very similar pattern.

14.4 Statistics of casualties and property losses

In Table 14.2 the following statistical data are tabulated:

1. Number of fires
2. Rate of fire fighter injuries
3. Rate of civilian injuries
4. Rate of civilian fatalities
5. Average estimated dollar loss.

The data cover apartment buildings as well as office buildings.

In Table 14.2 the general pattern is similar to the pattern for extent of fire damage, except that protected construction is more effective than detectors for all categories except for civilian casualties in offices. In addition, detectors in offices appear to have statistically no benefit as far as dollar loss is concerned.

It can certainly be said that, in general, it is better to have sprinklers than either detectors or protected construction. An important point to note is that none of these systems (sprinklers, detectors or protected construction) are 100% effective against any of the criteria examined. But whether this is due to less than 100% reliability or less than 100% efficacy, or both, remains an open question.

Table 14.2 Casualties and property dollar losses

	Number of fire safety system present	Protected construction	Detectors	Sprinklers
Apartments				
Fire fighter injuries	54.9	43.3	59.5	36.0
Civilian injuries	65.5	71.4	86.8	28.3
Civilian fatalities	9.4	7.4	8.7	2.6
Average estimated dollar loss	8451	5520	6808	3613
Number of fires	42,666	53,075	51,988	389
Offices				
Fire fighter injuries	88.4	57.0	57.1	53.6
Civilian injuries	14.2	16.6	13.1	11.9
Civilian fatalities	1.2	0.5	1.9	0.0
Average estimated dollar loss	25,832	16,388	38,614	8730
Number of fires	4026	4335	1069	168

Notes: Injuries and fatalities are per 1000 fires.
Dollar losses are in US dollars per fire.

However, the decision to install sprinklers cannot be made on the basis of their efficacy alone, in view of the fact that the installation of sprinklers is expensive. A comprehensive cost–benefit analysis should be carried out on the lines of Section 7.5. In carrying out such a cost–benefit analysis, it will be necessary to estimate the reliability of the sprinkler system. This can be done through the use of fault tree analysis (see Chapter 6). For an example of such an analysis see Section 17.3.2.

14.5 Conclusion

An important point to be drawn from the material presented above is that the use of sprinklers, detectors or protected construction (and presumably any other fire safety system, subsystem or component) does not guarantee that there will be no casualties and no property damage. More importantly, the only way to compare the effectiveness of these and other systems is to look at the outcomes of very large numbers of fires where they are present.

Rational design of fire safety systems for buildings requires an understanding of the relative effectiveness of the fire safety systems, subsystems and components that may be used in their construction. Based on the material presented above it appears that sprinklers are more beneficial than detectors and protected construction in limiting the extent of flame damage and that they are generally (but less clearly) more beneficial than the others in limiting fire fighter injuries, civilian fatalities and average estimated dollar loss. But a decision to install sprinklers must be based on a cost–benefit analysis.

15

Stochastic modelling of fire brigade response

15.1 Introduction

Fire brigade response is an essential component of the fire risk assessment process and so must be considered when the fire risk of a building is being determined.

In the adopted approach, the operation of the fire brigade was divided into separate components. They were then connected again through an event tree, making it possible to evaluate the overall probability of success.

A useful feature of the presented model is that the building fires where a fire unit is able to arrive early enough to have a good chance to save the target can be distinguished from those in which the fire safety depends completely on an automatic fire extinguishing system.

The work presented in this chapter is based on the article by Tillander and Keski-Rahkonen [67].

15.2 Operative time distributions

The *turnout time* is the length of time from the moment the fire brigade is notified until it leaves the fire station. The *response time* is the length of time from the moment the fire brigade is notified until the unit is at the fire scene. It is the sum of the turnout time and the travel time. The *operating time* is the length of time from the moment the fire brigade is notified until the fire unit returns to the fire station.

Curve fitting to the distributions of the turnout time, the response time and the operating time was carried out and it turned out that the gamma distribution seemed a plausible fit. The gamma distribution is a non-negative continuous distribution with density function

$$f(t) = \frac{1}{\Gamma(\alpha)\beta^{\alpha}} t^{\alpha-1} e^{\frac{1}{\beta}}, \tag{15.1}$$

where $\Gamma()$ is the gamma function defined in equation (3.187), and α and β are positive parameters. Its mean is $\alpha\beta$ and its standard deviation is $\beta\sqrt{\alpha}$. A typical shape of the gamma density function is illustrated in Figure 15.1.

Table 15.1 gives the mean and standard deviation of the turnout, response and operating time for the years 1994–1997.

All three distributions have a very long tail.

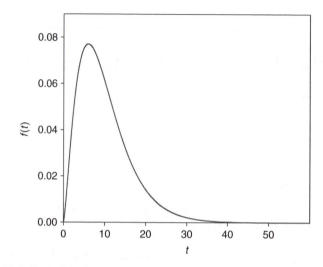

Figure 15.1 Typical density function of gamma distribution.

Table 15.1 Mean and standard deviation of fire brigade times

	Mean (min)	Standard deviation (min)
Turnout time	2.56	2.02
Response time	10.00	6.32
Operating time	76.00	55.14

15.3 A model for the travel time

Travel time is actually a more or less deterministic function of the distance between the fire station and the location of the fire. In a model proposed by Kolesar [41] it is assumed that the fire unit at first accelerates until it reaches its full cruising speed, which it maintains for a period and starts to decelerate as it approaches its destination. This process can be repeated several times during one drive because of road and traffic obstacles but this does not affect the mathematical shape of the model. According to Kolesar, the travel time t (in min) is given by

$$t = \begin{cases} a\sqrt{s} & \text{for } s \leq d, \\ bs + c & \text{for } s \geq d, \end{cases} \tag{15.2}$$

where s is the travel distance (in km) and a, b, c and d are parameters to be estimated from the data. On account of the laws of mechanics there must also be two additional requirements:

1. t must be a continuous function of s,
2. the first derivative of t with respect to s must be continuous.

A typical formula derived from data referring to rescue vehicles in Finland is

$$t = \begin{cases} 3.0\sqrt{s} & \text{for } s \leq 8.0, \\ 0.5s + 4.2 & \text{for } s \geq 8.0. \end{cases} \tag{15.3}$$

One way of evaluating the goodness of fit of the model to the data is to calculate the correlation (3.10.1) between the observations and the corresponding calculated values. The correlation turned out to be 0.84, which is reasonably strong.

Of course the travel time remains a random variable, but this is because the distance between the fire and the fire station is random.

15.4 The average number of simultaneous fires

The average number of simultaneous fires C in the area under study is given by

$$C = \lambda\tau, \tag{15.4}$$

where λ is the number of fires per unit time (the fire rate) and τ is the average operating time per fire. For example, in Helsinki, for building fires, the fire rate per hour per million inhabitants was 0.049, the average operating time 0.74 h and so the average number of simultaneous building fires was 0.036. For all fires, the average number of simultaneous alarms was 0.76.

Of course, the number of simultaneous fires varies with different periods of time. In particular, there is a significant diurnal variation, with the maximum around 6 p.m. to 7 p.m. and the minimum around 8 a.m. to 9 a.m.

15.5 The frequency of blockage

Blockage refers to the situation when all fire units are out when a fire is notified. The probability of blockage P_B can be assessed by using a formula originally obtained by Erlang in connection with telephone exchanges, where blocking also occurs when an incoming telephone call finds all lines busy. The formula is

$$P_B = \frac{\frac{C^S}{S!}}{\sum_{r=0}^{S} \frac{C^r}{r!}}, \tag{15.5}$$

where C is, as before, the average number of simultaneous fires and S is the number of available fire units.

Erlang's formula was developed assuming that the operating time distribution is negative exponential. As pointed out above, the operating time actually has a gamma distribution. But it can be proved that Erlang's formula still applies.

The population of Helsinki in 2000/2001 was 0.56 million. Thus the average number of simultaneous alarms was $0.76 \times 0.56 = 0.43$.

Figure 15.2 shows the dependence of the blockage probability on the number of fire units available. If the acceptable blocking probability is set between 10^{-4} and 10^{-5} the required number of fire units is five.

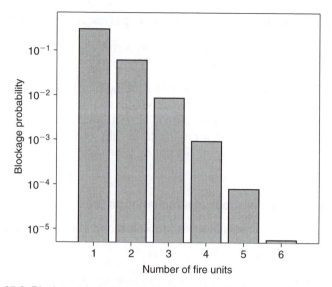

Figure 15.2 Blockage probability dependence on number of fire units.

More advanced theories of blockage have been developed. For references see Tillander *et al.* [67].

15.6 The event tree for fire extinction performance

Analysis of the success of fire extinction can be described by an *event tree* (see Chapter 6). The fire extinction mission is divided into four steps:

1. Departure
2. Whether or not the fire unit is delayed
3. Whether or not the location is found
4. Whether or not extinction succeeds.

The probabilities of a negative outcome at each step, conditional on the preceding ones, are represented by F_1, F_2, F_3 and F_4. Figure 15.3 illustrates the event tree for fire extinction.

The probability of successful fire extinction is given by

$$(1 - F_1)(1 - F_2)(1 - F_3)(1 - F_4). \tag{15.6}$$

15.7 A design example

The building used in the design example of this section is a large shopping centre (area over $10,000 \text{ m}^2$) constructed in Central Finland. The response time of the fire brigade was compared to the time scale set by a design fire described in Ref. [6]. This fire can be considered as one of the most severe design fires likely to be encountered in a shopping centre of this kind.

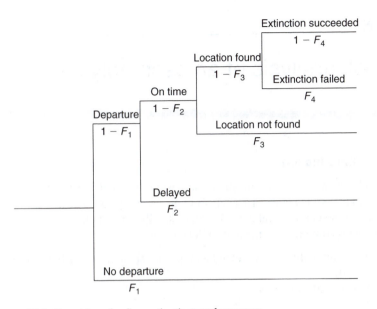

Figure 15.3 Event tree for fire extinction performance.

If the sprinklers do not operate, the critical rate of heat release is reached after 6.2 min. Assuming that there are five units available the probability of blockage F_1, calculated from formula (15.5), is 0.001. This is a very low probability and therefore blockage is not in this case a significant factor.

Given that a fire unit is available, the probability F_2 that it arrives on the scene on time is 0.30. According to an expert estimate, the probability of locating the seat of the fire is 0.95. Finally, the probability that fire extinction will fail turns out to be 0.20.

The probability of successful extinction of the fire can be calculated using equation (15.6):

$$(1 - F_1)(1 - F_2)(1 - F_3)(1 - F_4)$$
$$= (1 - 0.001) \times 0.30 \times 0.95 \times 0.80 = 0.23. \qquad (15.7)$$

This is a very low probability, and it indicates that if the sprinklers do not operate it is unlikely that the fire brigade will be able to control the fire. However, if the sprinklers operate, the fire will remain small and local, and will be easily extinguished. Therefore, in this case, acceptability of the fire safety design will depend on the reliability of the sprinkler system.

16

Risk analysis of an assembly hall

16.1 Introduction

In this chapter, we shall illustrate the design of a performance-based safety system by looking at the risk analysis of an assembly hall. The risk assessment will be based on the evaluation of the probability distribution of the available safe egress time G. Two methods will be used:

1. The β reliability index method and its corresponding probability of failure $\Phi(-\beta)$.
2. Monte Carlo simulation.

This analysis is based on the work of Magnusson *et al.* [44].

The basic variables of the problem are as follows:

A, floor area of the hall in m^2.
H, ceiling height in m.
N, density of hall occupants in persons per m^2.
G, available safe egress time.
S, time for the smoke to fill the lower part of the hall to 1.6 m above floor level.
D, detection time in s.
R, response and behaviour time prior to evacuation in s.
E, movement time to exit in s.
α, fire growth rate in kW/s^2.
F, specific flow capacity of a doorway in persons/s m.
W, exit door width in m.

The scenario event tree is shown in Figure 16.2 outlining the eight various outcome cases for functioning/not functioning fire alarms, sprinklers and emergency doors. However, in the rest of this chapter, we shall consider only scenario six: an automatic fire alarm is installed and operating; no sprinkler is installed (or operating).

The exit width will be taken as 4.8 m (all evacuation doors available) in the first instance.

16.2 Governing equations for scenario six

1. The theoretical available safe egress time G is given by the equation

$$G = S - D - R - E \qquad (16.1)$$

In addition, modelling uncertainties are represented by three extra variables: M_S, M_D and M_E, so that the final egress time G is given by

$$G = M_S S - M_D D - R - M_E E \qquad (16.2)$$

It is assumed that safety will be assured if $G \geq 0$.

2. Based on regression analysis of the CFAST computer fire model [11], the time S for the smoke to fill the lower part of the hall to 1.6 m above floor level is given by the equation

$$S = 1.67\alpha^{-0.26}H^{0.44}A^{0.54}. \qquad (16.3)$$

For further details see Example 1 in Section 3.14.

3. The detection time D is given by the equation

$$D = 5.36\alpha^{-0.478}H^{0.7}. \qquad (16.4)$$

4. The movement time to exit E is given by

$$E = \frac{NA}{FW} \qquad (16.5)$$

where F is the specific flow capacity of a doorway. It will be taken as 1.3 persons/s m.

16.3 Parameter distributions

A number of parameters are endowed with a probability distribution, as follows:

1. α: Uniform (see Section 3.8). $f(x) = 10.1$ for $0.001 \leq \alpha \leq 0.1$ with mean 0.05 and standard deviation 0.025.
2. H: Uniform. $f(x) = 0.11$ for $3 \leq H \leq 12$ with mean 7.5 and standard deviation 2.25.
3. A: Uniform. $f(x) = 0.001$ for $200 \leq A \leq 1200$ with mean 700 and standard deviation 250.
4. R: Lognormal (see Section 3.11.2) with mean 130 and standard deviation 120. It follows, using the formulae of Section 3.11.2, that $\ln(R)$ is normally distributed with mean 4.60 and standard deviation 0.785.
5. N: Triangular with parameters $(0.1, 0.8, 1.0)$ with mean 0.63 and standard deviation 0.40. The shape of the density function is given by Figure 16.1.
6. M_S: Normal with mean 1.35 and standard deviation 0.1.
7. M_D: Normal with mean 1.0 and standard deviation 0.2.
8. M_E: Normal with mean 1.0 and standard deviation 0.3.

There are thus eight basic variables and they are assumed to be independent. The limit state function can be written explicitly as

$$G = (1.67\alpha^{-0.26}H^{0.44}A^{0.54})M_S - (5.36\alpha^{-0.478}H^{0.7})M_D - R - \frac{NA}{FW}M_E.$$

$$(16.6)$$

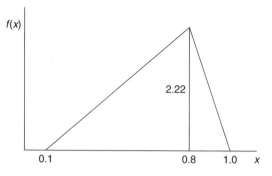

Figure 16.1 Density function of triangular distribution of *N*.

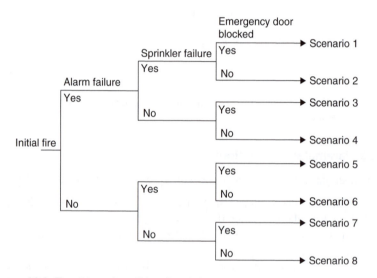

Figure 16.2 Event tree describing the eight scenarios.

16.4 Results

16.4.1 Beta reliability index

Using the methodology described in Chapter 4, the reliability index β and the corresponding probability of failure $p_f = \Phi(-\beta)$ were calculated for each scenario. For scenario six and a door width of 4.8 m the results were $\beta = 0.870$ and $p_f = 0.19$.

For the eight scenarios outlined in Figure 16.2, the range of β was 0.247–1.04, and the corresponding range of p_f was 0.402–0.149.

As the β values obtained are quite low, the effect of the door width on the β value was investigated. The calculation was performed on scenario six. The results for the β values and the probability of failure are shown in Figure 16.3. As the door width is increased to 20 m, the β value is increased to 1.18 and p_f

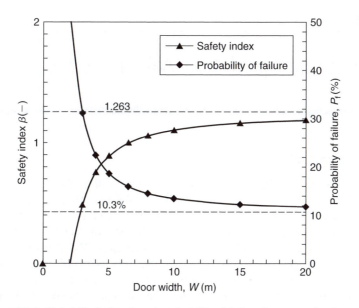

Figure 16.3 Reliability index β and probability of failure for scenario six.

decreases to 0.119. At the extreme, when $W \to \infty$, β reaches the value of 1.26 and the probability of failure decreases to 0.103.

16.4.2 Monte Carlo simulation

A simulation using random sampling was carried out, with a sample size of 10,000. The probability of failure was 0.1952, with a confidence interval (equation (5.4)) (0.1875, 0.2029), which contains the β reliability value, showing that the two methods are consistent for this analysis.

16.5 Discussion

The reliability of the proposed design is clearly much too low. A more detailed analysis (see Ref. [44]) shows that the sources of the large probability of failure are:

1. The large variance of the occupant response time R, which is clearly the controlling variable.
2. The fact that the results apply to a whole class of buildings with a large range of values for the height and the floor area design parameters.

Magnusson *et al.* state that this was a pilot investigation and that the distributions characterizing the random elements were chosen subjectively by a small group of persons. It is clearly possible that the distributions were chosen conservatively, i.e. by exaggerating the actual or real uncertainty. This could explain the fact that the calculated safety levels seem rather low.

17

The building at 140 William Street

17.1 Introduction

The case history of the 41-storey building at 140 William Street, Melbourne, Australia, detailed in Thomas *et al.* [66] vividly illustrates the potential benefits of risk-based fire safety design. There, compliance with the existing fire code would have required an extremely costly replacement of the inter-floor insulation. The problem was put in the hands of a team from BHP (Broken Hill Pty Ltd) Research Laboratories. They used a pioneering risk-to-life analysis based on stochastic modelling to compare the safety of a building complying with existing requirements and that of an alternative design based on improving the reliability of the sprinkler system. The latter turned out to provide greater safety at a vastly reduced cost and it was eventually adopted by the authorities.

In 1991, the existing building was a large freestanding building located centrally in the City of Melbourne. It had 41 levels, of which 37 could be occupied, and was used as an office building. See Figure 17.1 for an elevation of the building. All the columns and the bracing members were heavily encased in concrete and in the case of the external columns were further encapsulated with steel plate. The floors were supported on long-span castellated steel beams framing between primary beams and the perimeter of the building. A typical floor framing plan is shown in Figure 17.2. The floor slabs were of composite construction using lightweight concrete and steel decking. The soffit of the floor slabs and steel supporting beams were sprayed with asbestos-based fire protection material.

It was proposed that the building be refurbished and this included removal of the asbestos based fire protection material from the beams and from the soffit of the composite floor slabs. In terms of the requirements of the regulations then current, namely the Building Code of Australia (BCA) [19], the slab and beams would have to be protected with a suitable material.

In addition, the existing sprinkler system was only suitable for extra light hazard (sprinkler heads generally at 4.6 m centres) and there were no sprinklers in the ceiling spaces, whereas the regulations then current required a system suitable for ordinary hazards (sprinkler heads generally at 3.5 m centres) as well as sprinklers in the ceiling spaces.

However, the question whether these measures were really necessary for the building to be satisfactorily safe was raised with the relevant regulatory officer, the Building Surveyor of the City of Melbourne. He indicated that he would be sympathetic to rational argument as to why the additional measures were not required. But considering the magnitude and general implications associated

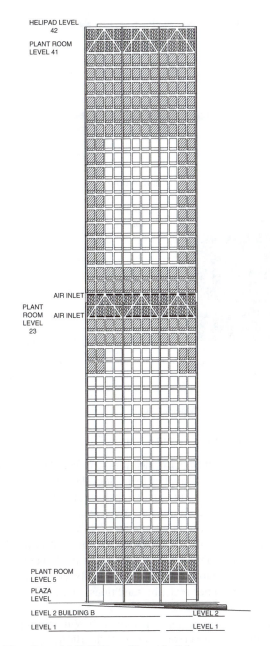

Figure 17.1 Elevation of 140 William Street building.

with an affirmative decision, would prefer to have the matter decided by the Building Referees of Victoria.

During 1990, submissions were made to the Referees, who responded by forming a specialist supervisory committee to review the proposed departures from the regulations and determine whether they should be accepted.

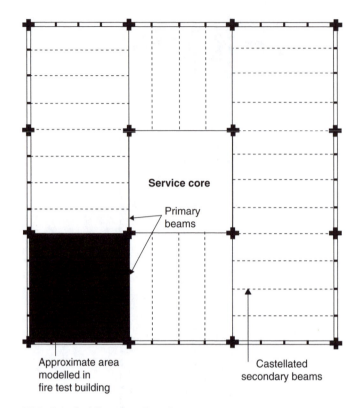

Figure 17.2 A typical floor framing plan.

The proposal put to the committee was as follows:

1. That a risk assessment be performed for the refurbished building, assuming no fire protection material applied to the floor slab and beams, and the current sprinkler system. It was later agreed that the risk assessment be based on the "Draft National Building Fire Safety Systems Code" as published in "Microeconomic Reform: Fire Regulation" [1].

2. That a series of fire tests be performed to provide data for the risk assessment model in relation to such matters as the likely nature of the fire, the performance of the current sprinkler system, the behaviour of the floor and castellated beams under real fire conditions, and the likely generation of smoke and toxic products.

3. That a risk assessment be performed for 140 William Street assuming it to conform to the minimum requirements of the BCA.

4. That if the risk-to-life obtained for the refurbished building described in item (1) was less than or equal to that obtained in item (3), then it would not be necessary to fire protect the floors or castellated beams in the refurbished building and it would not be necessary to alter the existing sprinkler spacing.

The intention was that an improved fire safety system be incorporated in the refurbished building that would more than compensate for any (if indeed there was any) lesser fire performance with consequent increased risk resulting from the deletion of the fire protection of the floor slabs and beams, and at a lower cost than the fire protection.

This proposal was accepted by the specialist committee subject to the following conditions:

1. That all aspects of the proposed tests be approved by the supervisory committee.
2. That the fire load in the test building be approved by the supervisory committee.
3. That the results of the risk assessment be independently reviewed by Prof. Vaughan Beck of the Victoria University of Technology.

Following extensive discussions with the supervisory committee the details of the test building and a programme of four tests were agreed upon.

The 140 William Street building was owned by the AMP Society, which provided financial support for the research programme. Lincoln Scott Australia Pty Ltd, the engineering consultants for the project, proposed the research programme and acted in a technical coordination capacity.

17.2 Basis for the risk assessment

17.2.1 The simplified risk assessment procedure

The Draft National Building Fire Safety Systems Code [1] provides detailed requirements for a "simplified" risk assessment that meets the requirements set in the previous section, based on the use of a **risk assessment model** that is outlined in great detail.

It states that the basis of the simplified risk assessment. *"shall be that the occupants of the building, fire brigade personnel and the occupants of adjoining buildings are alive unless they are in an enclosure which is subjected to untenable conditions"*.

The simplified risk assessment views the fire safety system as being made up of several **subsystems**, each of which is represented by a **submodel**. Thus

"the risk assessment model shall incorporate at least the following submodels:

- *Fire ignition and development*
- *Spread of smoke and fire products*
- *Flame spread*
- *Occupant communication and response*
- *Fire brigade communication, response and operations*
- *Occupant avoidance.*

The models used to determine the risk-to-life shall be appropriate to the application and shall be based on, or correlated with, test results that are relevant to the application."

For simplicity, the draft uses the concept of **design fires** and reduces these to three basic types:

> *"The risk assessment model, and the relevant submodels, shall consider the development and spread of fire and its effects throughout the building using at least the following fire scenarios in the enclosure of fire origin:*
>
> 1. *A smouldering fire*
> 2. *A non-flashover fire*
> 3. *A flashover fire.*
>
> *The risk assessment shall consider the effects of each of the above fire scenarios starting in each enclosure in the building."*

The risk assessment process involves:

1. The ability of the occupants to recognize the presence of a fire through the recognition of **cues**. The draft simplifies this process by defining specific occurrences in the fire development process as the "cues".
2. The occurrence of life-threatening situations. This is also simplified by defining **untenable conditions**.

The design of the subsystems is covered in the draft in the following terms:

> *"The design shall also identify the role of each subsystem and component in terms of the cues which they are involved in providing. Unless shown otherwise by an appropriate analysis, the probability of occurrence of the cues shall be the product of the probabilities of all the subsystems or components that are required to provide the cue."*
>
> *"The design shall also identify the role of each subsystem and component in terms of the untenable conditions which they are designed to prevent. Unless shown otherwise by an appropriate analysis, the time of occurrence of an event shall be assumed to occur at the least time of failure of any of the subsystems or components that may influence the event."*

For each fire scenario and each enclosure of fire origin, the consideration of fire initiation and development is required to include:

1. the estimation of the probability of the initiation of fire,
2. the probability of initiation of the various fire types,
3. the development of a fire and its effects,
4. the control of extinguishment of the fire by various means,

to estimate the time of occurrence and the probability of the occurrence at that time of untenable conditions in the enclosure of fire origin.

Similarly, consideration of the spread of smoke and toxic products and of the spread of flame is required to be in terms of:

1. the time of operation and the probability of operation at that time of each subsystem and/or component,
2. the time of failure and the probability of occurrence of failure at that time of each subsystem and/or component,

to estimate, for each fire scenario and for each enclosure of fire origin:

1. the time of occurrence,
2. the probability of occurrence at that time,

of untenable conditions in enclosures other than that of fire origin.

In addition, consideration of occupant communication and response is required for each fire scenario and each enclosure of fire origin. It is required to include at least the time of occurrence and the probability of occurrence at that time of two types of cues for the occupants of each enclosure:

1. cues that communicate information via audible, visual and tactile **alarms** to the occupants;
2. cues associated directly with the "initiation and development" of a design fire that can be detected by one or more of the senses (visual, auditory and olfactory) or by people alerting or warning others.

In considering the fire brigade communication and operation, the designer is required to estimate the time, probability of occurrence and effectiveness of fire brigade intervention, because this can be significantly influenced by the fire safety system in the building.

Finally, consideration of occupant avoidance is required to include at least the characteristics of the occupants of the building, the likely response of the occupants to the fire cues and their egress times in determining the time and probability of occupant response to cues – and from that the number of people exposed to untenable conditions in each enclosure.

17.2.2 Basis for the risk assessment on 140 William Street

Risk assessments were carried out on three buildings:

1. *Building E*: The existing building at 140 William Street (the *Existing* building).
2. *Building B*: A building similar in construction, geometry and layout to the existing building but with a minimum fire safety system which would satisfy the requirements of the BCA (a *BCA* building).
3. *Building R*: The proposed refurbished building (the *Refurbished* building) which is also similar in construction, geometry and layout to the existing building but with a modified fire safety system.

The assessment took the form of a comparison between the three buildings, addressing only the relevant differences between the three buildings in respect of the design of the relevant subsystems and components.

17.2.3 The design fires

The draft code allows for the use of experimental data as the basis for design and assessment. Fire tests conducted at the Broken Hill Pty Ltd Research, Melbourne Laboratories [65] provided the basis for much of the data, so that there was no explicit consideration of the design fires of the draft code other than the following:

1. The design fire termed a "smouldering fire" was found by a study carried out within the framework of the Warren Centre Project [23] to not present a significant threat to occupants who are awake and aware, as the occupants of a building such as this would be expected to be. Consequently it was not considered to be life threatening and thus was not considered further in the risk assessment.
2. The "non-flashover fire" was considered to be a non-spreading fire and so was not considered to be a threat to the occupants as they would become aware of its occurrence and be able to escape its effects. Consequently it was not considered further in the risk assessment. It is also the type of fire considered to occur if the sprinkler system operates and partially controls but does not extinguish the fire.
3. The "flashover fire" was considered to be the only appropriate design fire in this case. However, while in smaller enclosures it is understood that flashover may occur, it is assumed that in larger compartments flashover may not actually occur. In this situation it has been assumed that "partial flashover" may occur, i.e. locally there may be full fire involvement – the equivalent of flashover having occurred – but in other areas of the enclosure the fire may not yet have fully developed or the fire may have substantially burned the fuel and the fire has begun to drop in intensity. This situation was observed in the tests [65]. It was assumed that in this case the fire has the potential to spread throughout the whole building if a pathway exists.

17.2.4 Waiting time distributions

In the draft code [1] there is a requirement that for each event there be a time and a probability of occurrence at that time. In other words, the time of occurrence of the considered event, counted from some chosen origin, is to be treated as a *random variable*.

For example, if a smoke detector was placed in a room and the contents of the room set on fire, the time between the beginning of the fire and the activation of the smoke detector would be a random variable, which could be denoted by T. The considered random variable, which, because time is a continuous quantity, would be a continuous random variable would have a distribution function $F(t) = P(T \leq t)$ and a probability density function $f(t) = \mathrm{d}F(t)/\mathrm{d}t$. Such a random variable is known as a *waiting time*.

It is important to note, though, that waiting times differ from other random variables in that there is a non-zero probability that the event considered might

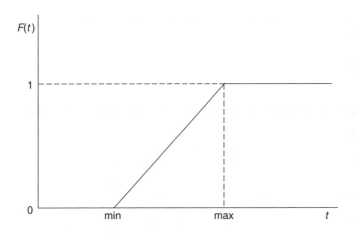

Figure 17.3 Distribution of waiting time.

not occur. For example, in the case of the smoke detector mentioned above, it is possible that the smoke detector may fail to activate altogether. There are several ways of dealing with this difficulty, but the simplest way is to associate with the event considered a probability p_0 of non-occurrence and then specify the probability distribution of the random variable *conditional* on the event actually occurring.

A difficulty that existed with the problem of specifying a distribution function for the various waiting times at the time of the risk assessment was the lack of sufficient data. However, from the data available, it was generally possible (with reasonable accuracy and confidence) to determine a range of times during which the event might occur (given that it would occur). It was then assumed that the distribution of the waiting time, given that the event occurred, was uniform between the maximum and the minimum times. The shape of the (conditional) distribution is shown in Figure 17.3.

17.3 Description of buildings E, B and R

The main purpose of this section is to describe in detail the differences between the existing building (E), the BCA building (B) and the refurbished building (R).

17.3.1 Smoke and Fire Barriers

Building E

A typical floor layout is given in Figure 17.4.

On each floor there are two fire doors (with automatic closers) between the lobby and the stairs.

Each floor of the building is a lightweight concrete composite slab incorporating Bondek sheeting. The soffits of the slabs are fire protected with fire spray material. The slabs are supported by steel beams which are also fire protected by fire spray.

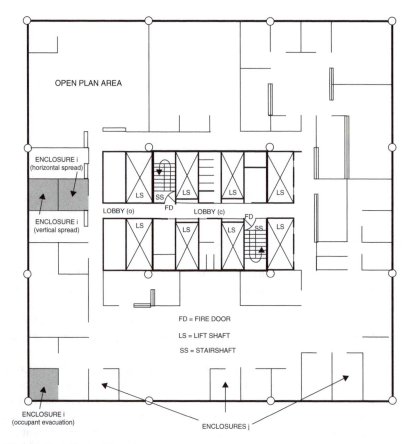

Figure 17.4 Typical floor layout.

Building B

Sandwich pressurization is provided and consequently (according to the BCA regulations) only one fire door (at the entrance to the stairshafts) is sufficient.

The soffits of the floors and the steel beams are fire protected to ensure that they will achieve a fire resistance level (FRL) of 120 min.

Building R

On each floor there are two fire doors between the lobby and the stairs. They are left permanently open with door closers which are automatically operated by smoke detectors or may be simply manually closed in the event of a fire.

The soffits of the floor slabs and the supporting steel beams are not fire sprayed. The performance of the unprotected floor slabs and steel beams has been demonstrated in the experimental programme of fire tests [65] in which the sprinklers were deliberately not operated. The floor system supported the applied loads without any sign of distress, even though gravity loads higher than required were applied to the floor, higher than reasonable fire loads were used and the floor slab was subjected to two unsprinklered burnout tests.

17.3.2 Automatic sprinklers

Building E

The existing sprinkler system is an extra light hazard system where the sprinkler heads are generally at 4.6 m centres. There are no ceiling sprinklers. A schematic diagram illustrating the essential details of the water supply system to the sprinklers is shown in Figure 17.5.

As shown in this diagram water is supplied from two separate mains. Within the building there are multiple risers with each riser supplying four floors with water. Each of the mains is then split by means of a trident to supply normal water, fire hose and sprinkler supplies. Over the life of the building there has only been one occasion when water was not available from **one** of the mains. Thus the building has never been without a main water supply.

Pressure is provided to the sprinkler system for storeys above the lowest eight storeys by electric pumps, backed by diesel pumps, operated from separate pressure switches. The electric power supply is exceptionally reliable. A fault tree diagram for diesel and electrical pump failures is given in Figure 17.6.

There is no external booster connection to the sprinkler system.

Building B

The required sprinkler system is a normal hazard system with the sprinkler heads at about 3.5 m centres. The piping is slightly larger (25 mm diameter instead of 20 mm).

The building has eight floors per sprinkler riser as opposed to four floors in building E. All the sprinkler stop valves and the main water supply valves are electronically monitored. The closing of a valve will sound an alarm at the fire indicator panel (FIP) and the fire brigade will automatically be notified.

The building has sprinklers in the ceiling space but at much wider spacing than that required below the ceiling.

The building has an external booster connection to the sprinkler system.

Building R

The existing sprinkler system is an extra light hazard system as in the existing building. The ability of this system to detect, control and extinguish a fire in both a small office and an open plan office has been demonstrated in the research programme [65].

As for the existing building, water is supplied to the building from two mains and there are four floors per riser. All the sprinkler stop valves and the main water supply valves are electronically monitored. The closing of a valve will sound an alarm at the FIP and the fire brigade will automatically be notified.

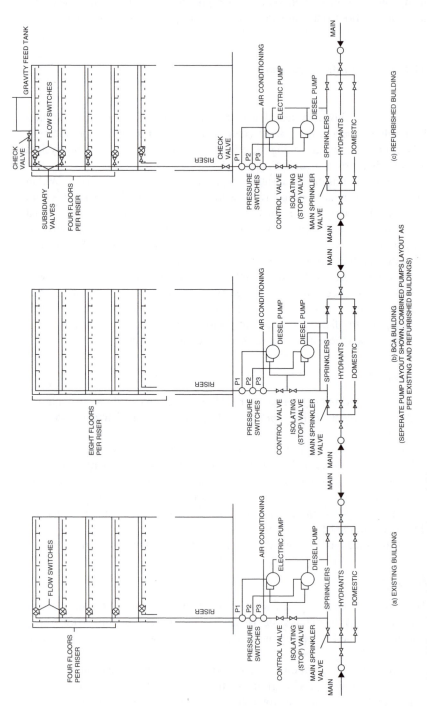

Figure 17.5 Water supply system to the sprinklers.

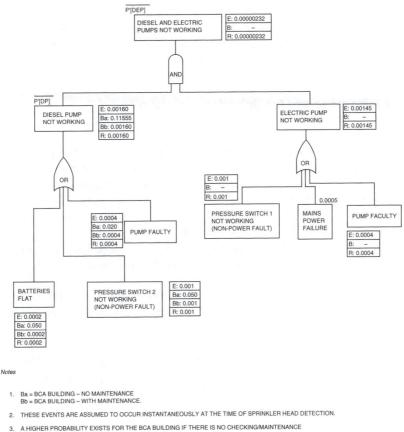

Figure 17.6 Fault tree for diesel and electrical pump failures.

Water may also be supplied to each sprinkler by means of a gravity supply system. The system can supply water and pressure to the sprinklers without requiring operation of either of the pumps in the building.

At each storey there is a subsidiary valve and a flow switch located in a fire-protected area. They are alarmed back to the FIP and to the fire brigade. The flow switches, if activated, will send an alarm to the FIP and activate the building alarm, but will not send an alarm to the fire brigade.

Solenoid valves are provided at every storey to allow remote end-of-line testing to be conducted on a weekly basis. Thus, the sprinkler system at every storey is to be tested weekly to see if there is water and whether the flow of water is indicated by the flow switch.

The provision of subsidiary valves at every storey means that only this valve, and not the stop valve on the riser, need be turned off to allow modification of the sprinkler system on a storey. This makes it unnecessary to use "spades", i.e. metal gates inserted into the sprinkler pipe to the storey in order to block water supply to that storey. If the "spade" is inadvertently left in place after completion of the modification and there is no regular testing of water flow the sprinklers on that storey will be inoperative.

17.3.3 Smoke and fire detection and alarm

Building E

Smoke detectors are located only within the return air ducts to the plant rooms. When a smoke detector is activated a signal is passed to the FIP. Similarly, if the water pressure drops (as measured by a pressure switch), possibly indicating the activation of the sprinklers, the air handling system is shut down and a signal is sent to the FIP. This results in an alarm being sent to the fire station and in the sounding of an audible fire alarm within the building.

In addition a direct brigade alarm (DBA) device is located on each group of risers and once water begins to flow through the system, an alarm is sent to the fire station.

The flow switches located at each storey, should they detect flow, result in activation of indicator lights on the FIP and in an audible alarm to the building occupants.

Breakglass alarms are located on each storey and their activation results in a signal to the fire station and an audible fire alarm throughout the building.

The FIP is powered by mains power but is backed up by a rechargeable battery. Should the battery become defective, this is indicated on the FIP.

Building B

Smoke detectors are located within the return air ducts at each storey. In addition, smoke detectors are provided in any equipment cupboard considered to present a possible fire hazard as well as adjacent to (and outside) the fire doors at the stairshaft at each storey level. A smoke detector is provided at the top of each stairshaft and adjacent to the fire door on the office side of the door.

Activation of any of the smoke detectors results in a signal being sent to the FIP and from there to the fire station and to the building alarm unit which sounds the audible fire alarm. Similarly, if the water pressure drops (as measured by a pressure switch), possibly indicating the activation of the sprinklers, an alarm is sent to both fire station and building occupants. Flow switches are not located on each storey.

DBA alarms to the fire station are provided as for the existing building.

Building R

The details are as for the BCA building except that smoke detectors are also included in the corridor adjacent to the additional fire doors separating the office area from the service shaft.

Activation of any smoke detector will result in a building alarm and notification of the fire brigade. Activation of a flow switch at any storey will result in an alarm at the FIP and an associated audible building alarm. No alarm is sent to the fire station as a result of flow switch activation.

17.3.4 Air handling equipment

Building E

Air handling is provided by plant equipment located at levels 5 and 6, level 23 and level 42. Should a fire in the building be detected by either the smoke detectors located in the return air ducts or by a water pressure drop (as measured by a pressure switch), possibly indicating the activation of the sprinklers, the air handling system is shut down. In the event of a fire, neither air pressurization nor sandwich pressurization is provided in the building.

Building B

Should a fire be detected at the FIP due to a water pressure drop (as measured by a pressure switch), possibly indicating the activation of the sprinklers, or any of the smoke detectors, the air handling system goes into a fire mode such that the stairways are pressurized and the storeys subject to sandwich pressurization.

Building R

Details are identical to the BCA building.

17.3.5 Building management

Building E

The existing building has a management team which is present 24 h day.

There is a well-established and practised evacuation policy for the building and a system of fire wardens to ensure correct and rapid egress from the building in the event of a fire. There is one fire warden per storey. The established evacuation procedure is that the storey on which the fire initiates be evacuated first, then the storeys above the fire floor, starting with the storey closest to the fire.

The building permits smoking on the premises except for areas specifically marked as non-smoking.

Maintenance of plant and equipment within the building is the responsibility of building management. As such, sprinkler, pump operation and the FIP are checked in accordance with AS1851.3 [47]. Modifications to the sprinklers to allow for occupancy changes were carried out by the one contractor over the life of the building. That contractor has a policy of not using "spades" when carrying out modifications.

Building B

The building is not required to be under full-time management and there is no established system of fire wardens and no definite evacuation procedure. Smoking is allowed in the building except where otherwise noted.

The BCA under Section E5.2 requires adequate maintenance of safety instal-
lations throughout the building but makes no specific reference to sprinklers or
to an associated standard. Requirements in Victoria [49] require maintenance of
sprinklers in accordance with AS1851.3 [47]. Both situations (i.e. maintenance
and no maintenance) have been considered in the risk analysis.

Building R

Full-time management team is present in the building. A system of fire wardens
and an established evacuation procedure as for the existing building will be
specified for the refurbished building.

Smoking is not permitted in the building.

Routine checking of the sprinkler/alarm system is undertaken in accordance
with AS1851.3 [47]. In addition, end-of-line testing of the sprinkler system is
undertaken on a weekly basis.

17.3.6 Lifts

Building E

All lifts in the existing building are classed as "emergency lifts". This means
that all lifts can be operated by firemen in the event of a fire for the purpose
of rapid transit to the storeys below the fire floor. These lifts could also be
operated by the fire brigade to achieve more rapid evacuation of the occupants
of the building should this be considered necessary. Four lifts provide transport
between the Plaza level (level 40) and office levels 7–14; an additional four lifts
provide transport between the Plaza and levels 14–22, while a further eight lifts
link levels 24–40 to the Plaza level.

Building B

Only one lift at any floor need be provided as an emergency lift. This means that
in the event of fire only one lift is available for the use of fire brigade personnel
to each floor.

Building R

All lifts are emergency lifts.

17.4 Risk assessment

17.4.1 Introduction

The objective of the risk assessment is to evaluate the risk-to-life in the three
buildings and to determine whether the refurbished building is at least as safe
as one which conforms to the BCA.

The model used can be visualized as an *event tree* (see Section 6.3). To evaluate the expected number of deaths, the probability of each branch is multiplied

by the number of deaths resulting from the events on that branch, and these are summed over all branches.

A grave difficulty in carrying out this task is the complexity of the model. There are over 60 choices to be made of whether or not certain events occur (e.g. the sprinklers do or do not operate). In addition, some outcomes will depend on the order in which events occur. So there are well in excess of $2^{60} \approx 10^{18}$ endpoints on the event tree. A full calculation is beyond the capabilities of commercial computers. However, most of the alternatives will represent such a low risk-to-life that they can be neglected.

A practical solution to this problem is to use Monte Carlo simulation (see Chapter 5). This method reduces the problem posed by complexity because it tends to choose the most probable events and so generally ignores the large proportion of the 2^{60} possible events which are highly improbable.

The number of trials required depends on two factors:

1. The complexity of the problem
2. How accurate an answer is required.

However, much of the data (the probabilities of events and the time of their occurrence) is not known accurately in this case. This limits the accuracy needed in the risk estimates. In any case, it was sufficient for the purpose of the assessment to be able to clearly rank the three buildings. Nevertheless, the complexity was sufficient to necessitate thousands, and for some submodels, even millions of trials.

17.4.2 The submodels

The following submodels were used in the risk assessment:

1. Smoke and flame management
2. Fire brigade communication and response
3. Occupant communication and response.

17.4.3 The fire scenario

The following fire scenario was assumed:

- The fire occurs when the building is normally occupied and the occupants are awake.
- The fire is initiated in a small office enclosure adjacent to both the external windows and the entrance to the core of the building.
- For the evacuation model the first occupant to evacuate is assumed to start from the corner of the building furthest from the stairs.
- Real office fires observed from fire tests of the test building [65] are used as design fires.

The event tree used, which is representative of event trees used throughout the risk assessment, is given in Figure 17.7.

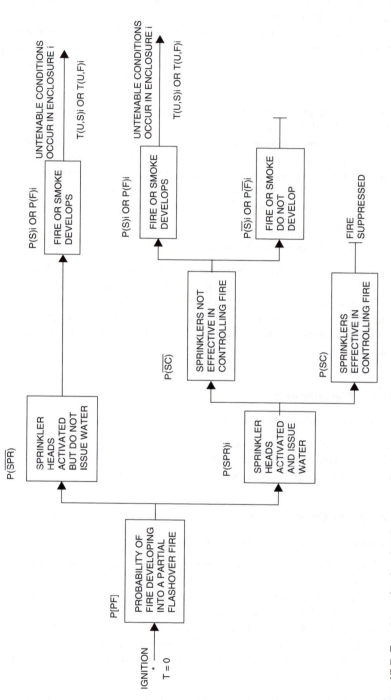

Figure 17.7 Event tree for the development of fire in an enclosure.

17.5 Results

Assessment of the risk-to-life in each building required estimates of:

1. The number of occupants of the building. Assuming 60 people on 34 storeys, there are 2040 occupants.
2. The rate of fire starts in office buildings per square metre per year. This is estimated in Ref. [23] for data on fire attendances in the Sydney (Australia) Central Business District (CBD) from 1986 to 1988 to be 8.9 per million square metre per year.
3. The proportion of fire starts that occur when occupants are in the building. This is estimated in Ref. [23] to be 0.54, assuming occupancy from 7 a.m. to 6 p.m.
4. The office area of each storey is taken to be $1520\,m^2$.
5. The probability that a fire develops into a partial-flashover type fire is estimated in Ref. [23] to be 0.2.

In summary, the number of expected deaths per year for the different buildings turns out to be

- E: 1.6×10^{-5}
- B: 9.0×10^{-5} without maintenance
  6.4×10^{-5} with maintenance
- R: 3.1×10^{-6} with extra fire doors
  3.3×10^{-6} without extra fire doors

Thus, the expected number of deaths per year is between 4 and 6 times higher in building B than in building E and between 20 and 30 times higher in building B than in building R.

17.6 Conclusion

The risk assessment was carried out to compare the estimated risk-to-life safety in the building as it existed at the time of the assessment, a similar building if built in accordance with the minimum requirements of the BCA, and the building after the proposed refurbishment.

The results of the risk assessment show that the risk-to-life safety in all the buildings is low, and that the refurbished building is substantially safer than the building satisfying the minimum requirements of the BCA. The calculated risk values compare with available "real-life" data. The relative safety of the three buildings examined is

- BCA building   safe
- Existing building   safer
- Refurbished building safest

This order remains valid whether or not maintenance to AS1851.3 [47] is carried out on the BCA building, and whether or not the outer fire doors are fitted in the refurbished building.

Thus it can be concluded that the proposed refurbished building more than adequately satisfies the fire safety objectives of the BCA.

Acknowledgements

Material from the following publications has been reproduced by permission of the respective copyright holders, as detailed below.

(A)

I.R. Thomas, I.D. Bennetts, S.L. Poon, and J.A. Sims. *The Effect of Fire in the Building at 140 William Street: A Risk Assessment.* BHP Research – Melbourne Laboratories, Rep. No. BHPR/ENG/92/044/SG2C, 1992.

The material reproduced consists of a 20 page summary of the publication, including six figures.

Copyright holder. Prof. I.R. Thomas, Centre for Environmental Safety and Risk Engineering, Victoria University of Technology, Australia.

(B)

Fire Code Reform Centre: Project 3 Report: Fire Resistance and Non-combustibility, 1996.
Fire Code Reform Centre: Project 4 Report, Part 1: Core Model and Residential Buildings, Modelling Phase, Volume 1: CESARE Risk: Fire Growth and Smoke Spread Models, 1998.
Fire Code Reform Centre: Project 4 Report, Part 1: Core Model and Residential Buildings, Modelling Phase, Volume 1: CESARE Human Behaviour Model, 1998.
I.D. Bennetts, K.W. Poh, S.L. Poon, I.R. Thomas, A.C. Lee, P.F. Beever, G.C. Ramsay, and G.R. Timms. Fire safety in shopping centres. Technical Report, Fire Code Research Program, Fire Code Reform Centre Limited, Australia, 1998.
Fire Code Reform Centre: Project 4 Report, Part 1: Core Model and Residential Buildings, Modelling Phase, CESARE-Risk: Evacuation Model – Computer Program and Results, 1998.

Chapter 12 of the book contains a summary of parts of item 2 and reproduces one figure of that item (Figure 7.2, p.91). Chapter 13 of the book contains a summary of parts of item 3 and reproduces four figures from that item, other items are just cited.

Copyright holder. Australian Building Codes Board, email: abcb.office@abcb.gov.au

(C)

R.L. Alpert and E.J. Ward. Evaluation of unsprinklered fire hazards. *Fire Safety Journal*, 7:127–143, 1984.

A.M. Hasofer and V.R. Beck. A stochastic model for compartment fires. *Fire safety Journal*, 28:207–225, 1997.

A.M. Hasofer and J. Qu. Response surface modelling of Monte Carlo fire data. *Fire Safety Journal*, 37:772–784, 2002.

S.E. Magnusson, H. Frantzich, and K. Harada. Fire safety design based on calculations: uncertainty analysis and safety verification. *Fire Safety Journal*, 27:305–334, 1996 (Figures 1 and 6 only).

G. Ramachandran. Statistical methods in risk evaluation. *Fire Safety Journal*, 2:125–145, 1979/1980.

S.A. Young and P. Clancy. Structural modelling of light-timbered framed walls in fire. *Fire Safety Journal*, 36:241–268, 2001.

Copyright holder: Elsevier Ltd.

(D)

P. Clancy. Advances in modelling heat transfer through wood framed walls in fire. *Fire and Materials*, 25(6):241–268, 2001.

Figures 22–24, p. 251.

Copyright holder: John Wiley and Sons Ltd, 1 Oldlands Way, Bognor Regis, West Sussex, PO22 9SA, England.

(E)

P. Clancy. A parametric study on the time-to-failure of wood framed walls in fire. *Fire Technology*, 38:243–269, 2002.

Figure 1, p. 244; Figure 4, p. 246; Figures 5 and 6, p. 247.

Copyright holder: Springer, Tiergartenstrasse 17, 69121 Heidelberg, Germany.

(F)

P. Clancy. Probability of failure with time for wood framed walls in real fire. *Journal of Fire Protection Engineering*, 12:197–223, 2002.

Figures 1, 2, 9, 10, 12, A1–A7, Table A1 (part of).

I.R. Thomas. Effectiveness of fire safety components and systems. *Journal of Fire Protection Engineering*, 12:63–78, 2002.

Part of Table 2, p. 73.

Copyright holder: Society for Fire Protection Engineers, 2002, by permission of Sage Publications Ltd, www.sagepub.co.uk

(G)

A.M. Hasofer and I.R. Thomas. Probability distribution of fire losses. In *Seventh International Symposium on Fire Safety Science, Boston*, June 2002.

Figures 1 and 2, p. 1066.

A.M. Hasofer, J. Qu, and I.R. Thomas. Analysing the time to untenable conditions. In *Interflam 2004, Edinburgh*, 2004.

Figure 10, p. 1059; Figure 11, p. 1060; Figure 13, p. 1061.

Copyright holder.   Interscience Communications Ltd.

(H)

G. Ramachandran. *The Economics of Fire Protection.* E&FN Spon, London, 1998. Cost benefit analysis of sprinklers, pp. 90, 94, 95.

Copyright holder.   E&FN Spon, 11, New Fetter Lane, London, EC4A IAP, UK.

(I)

Kati Tillander. Utilisation of statistics to assess fire risks in buildings. Draft 17.3.2004. Technical Research Centre of Finland, 2004.

Parts of Chapter 5 (including Tables 5 and 7) summarized in Chapter 8 of the book. Parts of Chapter 6 (including a modified form of Figure 28) summarized in Chapter 15 of the book.

Copyright holder.   VTT Information Service, P.O. Box 2000 (Vuorimiehentie 5A) FI-02044 VTT, Finland.

Bibliography

[1] *Draft National Building Fire Safety Systems Code*. Building Regulation Review Task Force, Canberra, ACT, Australia, 1991.

[2] National Fire Protection Association. Recommended practice for smoke-control systems. Technical Report, National Fire Protection Association, Quincy, MA, USA, 1993.

[3] National fire incident reporting system. Technical Report, Federal Emergency Management Agency, USA, Annual.

[4] K. Abe. The behaviour of survivors and victims in a Japanese nightclub fire: a descriptive research note. *Mass Emergencies*, 1:119–124, 1976.

[5] R.L. Alpert and E.J. Ward. Evaluation of unsprinklered fire hazards. *Fire Safety Journal*, 7:127–143, 1984.

[6] I.D. Bennetts, K.W. Poh, S.L. Poon, I.R. Thomas, A.C. Lee, P.F. Beever, G.C. Ramsay, and G.R. Timms. Fire safety in shopping centres. Technical Report, Fire Code Research Program, Fire Code Reform Centre, Australia, 1998.

[7] V. Babrauskas. Technical Note 1103. Technical Report, National Bureau of Standards, Washington, DC, USA, 1979.

[8] J.L. Bryan. Smoke as a determinant of human behaviour in fire situations (project people). Technical Report No. NBS-GCR-77-94, National Bureau of Standards, Washington, DC, USA, 1977.

[9] J.L. Bryan. Behaviour in the MGM Grand Hotel fire. *Fire Journal*, 76(2):37–48, 1982.

[10] J.L. Bryan. Human behaviour in the Westchase Hilton Hotel fire. *Fire Journal*, 77(4):78–85, 1983.

[11] R.W. Bukowski, R.D. Peacock, W.W. Jones, and C.L. Forney. *Technical Reference Guide for the HAZARD I Fire Hazard Assessment Method*. NIST, Gaithersburg, MD, USA, 1989.

[12] D. Canter, J. Breaux, and J. Sime. *Domestic, Multiple Occupancy and Hospital Fires*, pp. 117–136. John Wiley and Sons, Chichester, UK, 1980.

[13] Fire Code Reform Centre. *Project 3 Report: Fire Resistance and Non-combustibility*, 1996.

[14] Fire Code Reform Centre. *Project 4 Report: CESARE Human Behaviour Model*, 1998.

[15] P. Clancy. Advances in modelling heat transfer through wood framed walls in fire. *Fire and Materials*, 25(6):241–268, 2001.

[16] P. Clancy. A parametric study on the time-to-failure of wood framed walls in fire. *Fire Technology*, 38:243–269, 2002.

[17] P. Clancy. Probability of failure with time for wood framed walls in real fire. *Journal of Fire Protection Engineering*, 12:197–223, 2002.

[18] J. Cooper and D. Yung. NRCC – CNRC fire growth model for apartment buildings. Technical Report, Internal Report No. 734, Institute for Research in Construction, 1997.

[19] Australian Uniform Building Regulations Coordinating Council. *Building Code of Australia*. Department of Industry, Technology and Commerce, Canberra, ACT, Australia, 1988.

[20] J.L. Devore. *Probability and Statistics for Engineering and the Sciences*. Brooks/Cole Publishing Company, Pacific Grove, CA, 1991.

[21] I. Donald and D. Canter. *Behavioural Aspects of the King's Cross Fire Disaster*, pp. 15–30. David Fulton, London, 1990.

[22] F. Casciati and J.B. Roberts, editors. *Mathematical Models for Structural Reliability Analysis*. CRC Mathematical Modelling Series, 1996.

[23] The Warren Centre for Advanced Engineering. *Fire Safety and Engineering International Symposium Papers*. University of Sydney, Sydney, Australia, 1989.

[24] R. Friedman. A survey of knowledge about idealized fire spread over surfaces. *Fire Research Abstracts and Reviews*, 10:1–8, 1968.

[25] G.M. Haber. Human behaviour in fire depending on types of occupancy: health care, penal and leisure time activities. In Levin, B.M. and Paulsen, R.L., editors, *Proceedings of the 2nd International Seminar on Human Behaviour in Fire Emergencies*, October–November 1978, pp. 147–158.

[26] W.M. Haessler. *The Extinguishment of Fire*. National Fire Protection Association, Quincy, MA, USA, 1974.

[27] A.M. Hasofer. *Non-parametric Estimation of Failure Probabilities*, Chapter 4, pp. 195–226. In F. Casciati and J.B. Roberts [22], 1996.

[28] A.M. Hasofer and V.R. Beck. A stochastic model for compartment fires. *Fire Safety Journal*, 28:207–225, 1997.

[29] A.M. Hasofer and V.R. Beck. The probability of death in the room of fire origin: an engineering formula. *Journal of Fire Protection Engineering*, 10(4):19–28, 2000.

[30] A.M. Hasofer and J. Qu. Response surface modelling of Monte Carlo fire data. *Fire Safety Journal*, 37:772–784, 2002.

[31] A.M. Hasofer and I.R. Thomas. Probability distribution of fire losses. *Proceedings of the 7th International Symposium on Fire Safety Science*, Boston, June 2002.

[32] A.M. Hasofer, J. Qu, and I.R. Thomas. Analysing the time to untenable conditions. *Interflam 2004*, Edinburgh, 2004.

[33] A.M. Hasofer and I. Thomas. Analysis of fatalities and injuries in building fire statistics. *Fire Safety Journal*, 41:2–14, 2006.

[34] Y. He. CESARE-Risk: Fire growth and smoke spread models. Technical Report Vol. 1, Fire Code Reform Centre, Sydney, NSW Australia, 1998.

[35] Y. He and V.R. Beck. A computer model for smoke spread in multi-storey buildings. In Chan, S.H., editor, *Proceedings of the 8th International Symposium on Transport Phenomena in Combustion*, pp. 713–723. Taylor and Francis, San Francisco, 1995.

[36] H.E. Roland and B. Moriarty. *System Safety Engineering and Management*. 2nd edition, John Wiley and Sons, Hoboken, NJ, U.S.A., 1990.

[37] A. Hokugo, D. Yung, and G.V. Hadjisophocleous. Experiments to validate the NRCC smoke movement model for fire risk-cost assessment. *Proceedings of the 4th International Symposium of Fire Safety Science*, Canada, 1994, pp. 805–816.

[38] I.D. Bennetts, K.W. Poh, and I.R. Thomas. A framework for fire engineering design. *Australian Journal of Structural Engineering*, SE3(1&2):9–22, 2000.

[39] J.R. Benjamin and C.A. Cornell. *Probability, Statistics and Decision for Civil Engineers*. McGraw-Hill, New York, 1970.

[40] M.G. Kendall. *The Advanced Theory of Statistics*, Vol. 1. Charles Griffin and Co. Ltd, London, 1952.

[41] P. Kolesar. A model for predicting average fire company travel times. Technical Report R-1624-NYC, New York City-Rand Institute, 1975.

[42] T.T. Lie. *Structural Fire Protection: Manuals and Reports on Engineering Practice No.78*. American Society of Civil Engineers, Structural Division, 1992.

[43] S.E. Magnusson, H. Frantzich, and K. Harada. Fire safety design based on calculations: uncertainty analysis and safety verification. Technical Report LUTVDG/TVBB3078 SE, Department of Fire Safety Engineering, Lund University, Lund, Sweden, 1995.

[44] S.E. Magnusson, H. Frantzich, and K. Harada. Fire safety design based on calculations: uncertainty analysis and safety verification. *Fire Safety Journal*, 27:305–334, 1996.

[45] *Mathcad User's Guide*. Mathsoft, Inc., 101 Main Street, Cambridge, MA 02142, USA, 1997.

[46] D. Odigie. *The Investigation of Fire Hazards in Buildings Using Stochastic Modelling*. PhD Thesis, Victoria University, Melbourne, Vic. Australia, 2000.

[47] Standards Association of Australia. *Automatic Fire Sprinkler Systems, AS1851.3*, 1985.

[48] Standards Association of Australia. *Methods for Fire Tests on Building Materials, Components and Structures, Part 4: Fire Resistance Tests of Elements of Building Construction*, 1997.

[49] Victoria Building (Building Code of Australia) Regulations. *Statutory Rule No. 26*, 1991.

[50] Draft British Code of Practice. *The Application of Fire Safety Engineering Principles to Fire Safety in Buildings*, BSI, 1994.

[51] D. Purser. Toxicity assessment of combustion products. In *SPFE Handbook of Fire Protection Engineering*, Second Edition, pp. 2.85–2.146. National Fire Protection Association, Quincy, MA, USA, 1995.

[52] J. Rahikainen and O. Keski-Rahkonen. Determination of ignition frequency of fire in different premises in Finland. *Fire Engineers Journal*, 58(197):33–37, 1998.

[53] G. Ramachandran. Statistical methods in risk evaluation. *Fire Safety Journal*, 2:125–145, 1979/1980.

[54] G. Ramachandran. *The Economics of Fire Protection*. E&FN Spon, London, 1998.

[55] F.E. Rogers. Fire losses and the effect of sprinkler protection of buildings in a variety of industries and trades. Technical Report, Building Research Establishment, Fire Research Station, Borehamwood, Hertford shire, WD6 2BL, UK, 1977.

[56] R. Rutstein and R.A. Cooke. The value of fire protection in buildings. Technical report, Fire Research Report No. 16/78. Home Office Scientific Advisory Branch, London, 1979.

[57] L.A. Sanabria and S. Li. CESARE-Risk: evacuation model – computer program and results, Report of FCRC Project 4. Technical Report, Victoria University of Technology and Fire Reform Centre Ltd, Australia, 1998.

[58] J. Scanlon. Human behavior in a fatal apartment fire. *Fire Journal*, 73(3):122–123, 1979.

[59] T.T. Soong. *Random Differential Equations in Science and Engineering*. Academic Press, New York and London, 1973.

[60] R.D. Stewart, J.E. Peterson, T.N. Fisher, M.J. Hosko, E.D. Baretta, H.C. Dodd, and A.A. Hermann. Experimental human exposure to high concentrations of carbon monoxide. *Archives of Environmental Health*, 26:1–7, 1973.

[61] J.A. Swartz. Experimental human exposure to high concentrations of carbon monoxide. Human behavior in the Beverly Hills fire. *Fire Journal*, 74(3):73–74, 1979.

[62] H. Takeda and D.Yung. Simplified growth models for risk-cost assessment in apartment buildings. *Journal of Fire Protection Engineering*, 4(2):53–66, 1992.

[63] R.A. Thisted. *Elements of Statistical Computing*. Chapman and Hall, New York, 1988.

[64] I.R. Thomas. Effectiveness of fire safety components and systems. *Journal of Fire Protection Engineering*, 12:63–78, 2002.

[65] I.R. Thomas, I.D. Bennetts, P. Dayawanssa, D.J. Proe, and R.R. Lewins. *Fire Tests of the 140 William Street Office Building*. BHP Research – Melbourne Laboratories, Report No. BHPR/ENG/92/043/SG2C, 1992.

[66] I.R. Thomas, I.D. Bennetts, S.L. Poon, and J.A. Sims. *The Effect of Fire in the Building at 140 William Street: A Risk Assessment*. BHP Research – Melbourne Laboratories, Report No. BHPR/ENG/92/044/SG2C, 1992.

[67] K. Tillander and O. Keski-Rahkonen. The influence of fire department intervention to the fire safety of a building assessed using fire risk analysis. *Proceedings of the 3rd International Conference on Performance-Based Codes and Fire Safety Design Methods*, pp. 247–256. *Society of Fire Protection Engineers*, Bethesda, MD, USA, 2000.

[68] K. Tillander and O. Keski-Rahkonen. The ignition frequency of structural fires in Finland 1996–1999. *Proceedings of 7th International Symposium on Fire Safety Science*, Worcester, MA, USA, 2002, pp. 1051–1062.

[69] W.D. Walton and P.H. Thomas. Estimating temperatures in compartment fires. *SPFE Handbook of Fire Protection Engineering*, First Edition, pp. 2.16–2.32. National Fire Protection Association, Quincy, MA, USA, 1990.

[70] P.G. Wood. The behaviour of people in fires. Technical Report, Fire Research Note No. 953, BRE Fire Research Studies, Department of Environment and Fire Offices Committee, Joint Fire Research Organization, Borehamwood, UK, 1972.

[71] S.A. Young and P. Clancy. Structural modelling of light-timbered framed walls in fire. *Fire Safety Journal*, 36:241–268, 2001.

Index

Page numbers in *italics* refer to figures.